T0291040

Design of Hybrid Structures

Well-designed hybrid structures can combine different performance strengths of materials. This guide focuses on design approaches for concrete structures reinforced in an unconventional way by steel profiles. It explains force transfer mechanisms of steel profiles and concrete interfaces, and an analysis of the characteristics of hybrid structures, including slender components. Several types of hybrid designs are addressed: walls and columns with several embedded steel profiles, connections strengthened by steel profiles between steel and composite or reinforced concrete components, including the specific case of shear keys connecting deep beams or flat slabs to columns. The transition zones in partly reinforced concrete and partly composite columns are also covered.

Design of Hybrid Structures draws on the European SMARTCOCO research project of experimentation and numerical modelling, giving practical guidance for designers and introducing the subject to researchers and graduate students.

Design of Hybrid Structures

Where Steel Profiles Meet Concrete

Edited by
André Plumier and Hervé Degée

CRC Press
Taylor & Francis Group
Boca Raton London New York

CRC Press is an imprint of the
Taylor & Francis Group, an **informa** business

Cover image: Getty

First edition published 2023
by CRC Press
4 Park Square, Milton Park, Abingdon, Oxon, OX14 4RN

and by CRC Press
6000 Broken Sound Parkway NW, Suite 300, Boca Raton, FL 33487-2742

British Library Cataloguing-in-Publication Data
A catalogue record for this book is available from the British Library

---

Library of Congress Cataloging-in-Publication Data

Names: Plumier, André, editor. | Degée, Hervé, editor.
Title: Design of hybrid structures : where steel profiles meet concrete /
edited by André Plumier and Hervé Degée.
Description: First edition. | Abingdon, Oxon ; Boca Raton : CRC Press,
[2023] | Includes bibliographical references and index.
Identifiers: LCCN 2022051451 | ISBN 9780367712075 (hbk) | ISBN
9780367712082 (pbk) | ISBN 9781003149811 (ebk)
Subjects: LCSH: Composite construction. | Building, Iron and steel. |
Concrete construction. | Structural design.
Classification: LCC TA664 .D47 2023 | DDC 624.1/821--dc23/eng/20230103
LC record available at https://lccn.loc.gov/2022051451

ISBN: 978-0-367-71207-5 (hbk)
ISBN: 978-0-367-71208-2 (pbk)
ISBN: 978-1-003-14981-1 (ebk)

DOI: 10.1201/9781003149811

Typeset in Sabon
by Deanta Global Publishing Services, Chennai, India

Contents

Editors

André Plumier completed his PhD in the area of welding residual stresses and their effect on stability. He has participated in 20 research projects in earthquake engineering and hybrid structures. He conceived the dogbone, a specification for the ductility of steel structures. He contributed to developing Eurocode 8, a European standard for seismic design (2004) and to its revision of which he now also officially contributes. He provides support to projects involving hybrid details and seismic assessment and retrofitting, and he has also given seminars on seismic design.

Hervé Degée started as a research associate at the University of Liège, Belgium, working on stability issues in steel and steel-concrete structures in normal and seismic conditions. Since 2014, he has been a professor at Hasselt University, Belgium, where he heads the Structural Testing Laboratory (ACB²). He has participated in about 20 European research projects and is very active in the standardization at national and European levels, being a member of various working groups for Eurocodes 3, 4, 6 and 8.

Pierre Mengeot is a design project manager at BESIX, having skills in design, analysis and execution of civil engineering work and with 17 years of experience in world-wide projects such as marine works, quay walls in Africa and Oman, tall buildings in Azerbaijan and other civil structures, such as a posttensioned concrete bridge in Rotterdam, the Netherlands, which was built by the incremental launching method.

Rajarshi Das completed his PhD on innovative hybrid coupled wall structures with laser cut open-to-circular hollow section connections at Hasselt University, Belgium. He is currently working as a doctoral assistant and is a part-time teaching assistant at Hasselt University. He has previously participated in three European research projects and is currently involved in two EU-RFCS research projects as well as other internal projects.

Hugues Somja worked at Greisch Engineering for 14 years before joining the structural engineering research group at INSA Rennes. He has participated in five European research projects in earthquake engineering and steel-concrete structures. He has been active for ten years in the development of innovative concrete-steel systems for simple buildings in France in collaboration with industrial partners through the LabCom ANR B-HYBRID.

Pisey Keo completed his PhD in computational modelling of hybrid columns at INSA Rennes. He developed a design guide as well as a design tool for stability design of hybrid columns. He worked at LabCom ANR B-HYBRID for two years, extending his research on hybrid structures. He then spent three years at CTICM (Centre Technique Industriel de la Construction Métallique) researching composite steel-concrete structures in fire situations. In 2020, he became an associate professor at INSA Rennes, France.

Ahmed Elghazouli is a professor and head of structural engineering at Imperial College London, UK. He has over 30 years' experience in structural engineering research and practice. His main interests are related to the behaviour and design of structures under extreme loads. He has published widely in related areas and has contributed extensively to Eurocode development activities. He is a fellow of the Royal Academy of Engineering, the Institution of Civil Engineers and the Institute of Structural Engineers in the UK.

Dan Bompa is a chartered engineer and lecturer in structural engineering at the University of Surrey, UK. He is an academic visitor at Imperial College London, where he held a post-doc post for over six years. His research interests include the response and design of hybrid systems, and the performance and modelling of innovative construction materials. He is a member of the Institution of Structural Engineers (UK), a European Engineer (EUR ING) and a fellow of Advance HE (formerly HEA).

Contributors

Pierre Mengeot
BESIX Group
Brussels, Belgium

Rajarshi Das
Hasselt University
Hasselt, Belgium

Hugues Somja
INSA Rennes
Rennes, France

Pisey Keo
INSA Rennes
Rennes, France

Ahmed Elghazouli
Imperial College London
London, England

Dan Bompa
University of Surrey
Guildford, England

Chapter 1

Hybrid structures in the real world

Pierre Mengeot

CONTENTS

1.1 INTRODUCTION

Concrete and steel are a good "match" for the following reasons: they have nearly the same thermal dilatation coefficient (to ensure compatibility of deformation), and a homogeneous elevation of temperature does not generate significant internal stresses; steel embedded in concrete's alkaline state constitutes good protection against corrosion; concrete ensures good fire resistance; and steel has good adherence to concrete.

During the detailed design of iconic structures, the classic use of reinforced concrete structures as developed by Joseph Monier in the 19th century frequently fails to meet client wishes and architectural requirements. Two of the main disadvantages of reinforced concrete are its significant self-weight and its limited tension capacity, which were already partly mitigated at the beginning of the 20th century by the invention of the concept of prestressed concrete.

Hybrid structures are composite steel-concrete structures that are not the classic reinforced concrete structures as considered in Eurocode 2 but are also not composite structures in the sense of Eurocode 4. Hybrid structures are reinforced concrete structures in which some components are steel, composite or concrete, reinforced by means of several embedded steel parts. Such types of structures do not need to include the continuous steel skeleton which is the only reference in Eurocode 4.

The use of hybrid structures is becoming more of a necessity to reduce the dimensions of the structural components and to ensure good execution quality by avoiding the on-site congestion that can occur in processes of

DOI: 10.1201/9781003149811-1
                                                                           1

reinforcement. However, until recently, there were gaps in the knowledge of many components of hybrid structures, and questions were frequently raised by companies about their design that could not easily be answered. Solving issues was an opportunity for close interaction between designers and researchers in universities. Safe design solutions were found, but they evidenced the need for better knowledge of the design of hybrid structures. Therefore, the University of Liege took the initiative in 2012 to set up a research proposal covering several aspects of hybrid structures. The project was named SmartCoCo, for Smart Composite Components, and it was created in collaboration with Imperial College (UK), INSA Rennes (France), ARCELOR MITTAL (Luxemburg) and BESIX (Belgium). It was funded by RFCS, the Research Fund for Coal and Steel of the European Union. The research activities covered different types of hybrid situations, which are those covered by this book. A complete set of reports was produced as follows:

- An up-to-date report that gathered the information available before a complete definition of the research activity, presented in Plumier (2016).
- A generic design approach report that provided tentative suggestions to design test specimens.
- Specimens design reports and test reports.
- A calibration report, in which the test results were used to assess the validity of the tentative design methods.
- A design guide that gave only the practical indications for design based on the research.
- An executive summary report (SMARTCOCO, 2017); this is available on the internet, while the detailed reports listed above are available on request to the authors of this book.

This book is based on the work developed in the SmartCoCo project, with further consideration, analysis and efforts for clearer presentation in comparison to the reports produced in the research project. This book presents practical design approaches for concrete structures reinforced in an unconventional way by steel profiles.

1.2 PRACTICAL APPLICATION OF HYBRID STRUCTURES AND CONTEXT OF THE FOLLOWING CHAPTERS

In marine-piled concrete structures, prefab components are mandatory, as scaffolding cannot be used above water. Prefab planks as shown in Figure 1.1a are typically temporarily placed on top of steel piles. The temporary connection consists of a steel profile protruding from the precast planks as illustrated by Figure 1.1b.

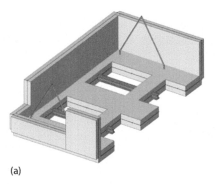
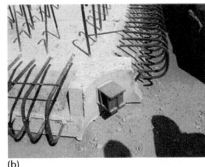

(a)   (b)

Figure 1.1 Precast plank. (a): 3D model precast plank. (b): Precast plank – detail protruding steel profile

The steel profile should be able to resist the self-weight of the precast plank, the steel cage added on top and the weight of the fresh concrete, all elements which together form, for instance, a monolithic mooring/berthing dolphin or marine platform. The challenge is not the definition of the steel profile, it is the transmission of force from steel to reinforced concrete; the steel profile will often be a welded built-up section because the precast plank thickness should be limited to reduce its self-weight. The transfer of force from the steel profile to the concrete can be done according to two mechanisms as illustrated by Option 1 and Option 2 in Figure 1.2. Both are statically valid, and each will transfer a part of the action effect F into the structure in proportion to the relative stiffness of the two mechanisms.

Chapter 2 of this book details the problem of force transmission between concrete and embedded profiles, a situation in which there are two unknowns: how to combine the resistance provided by bond, by friction and by connectors, which can be studs and/or plate bearings; and how to reinforce concrete in the transition zones between classical reinforced concrete zones and composite zones in order to avoid local damage. In high-rise buildings, the use of composite components is often a necessity to minimize the structural component size and to ensure the vertical and lateral stability of the structure. The starting point of the design process is commonly with classic reinforced concrete for economical reasons. However, due to specific architectural or client requirements, the maximum dimensions of vertical load-bearing components are often defined before they are designed, which leads to the use of high concrete grades and/or additional embedded steel profiles.

Chapter 3 of this book gives recommendations for the design of hybrid structures based on linear and non-linear analysis. It presents a simplified method for analyzing slender hybrid components subjected to an axial force combined with a uniaxial bending moment and the dedicated software developed during the SmartCoCo research project.

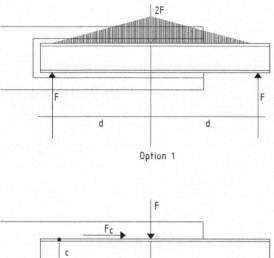

Figure I.2 Transfer of force from steel profile to concrete – two mechanisms

Chapter 4 explains how to calculate the action effects and the resistance of walls and columns with several embedded steel profiles. It also gives recommendations concerning the reinforcement details which are necessary for the design of such components. Advantages such as ductility, smaller reinforcement congestion, shear stiffness and resistance to in-plane effects are explained in this chapter.

Figure 1.3 shows a real-world example of a concrete column for which several embedded steel profiles were required in a few lower levels of a 54-floor residential tower. This local strengthening was necessary because of the great normal force combined with higher column slenderness due to increased story height in the podium or lower stories.

For use and stability reasons, the width of the bottom part or podium of a high-rise building and its foundations are often wider than the width of the top part of the building. This implies a diffusion of the vertical force by diagonal compression struts starting at the transition between the podium and the tower, as illustrated in Figure 1.4, which shows one main core wall of the 54-floor tower cited above and the diffusion of forces in its bottom part. The level of stress of such inclined struts depends on the stiffness of the foundation and on the soil-structure interaction set forward by the

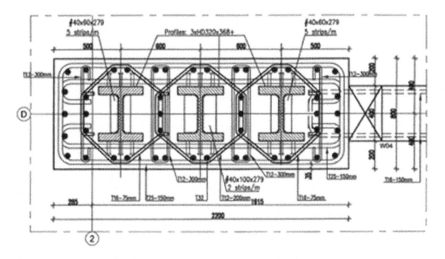

Figure 1.3 Concrete column with three embedded steel profiles

analysis. Significant potential differential settlement between the central part and the podium part of the foundation may imply a larger diffusion of force and therefore high local stress in the diagonal compression struts.

Due to the presence of an opening next to the inclined strut, a classical reinforced concrete structure was not an option anymore. Making the

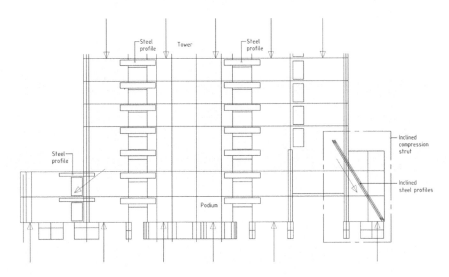

Figure 1.4 50-floor tower – elevation of main core walls in podium

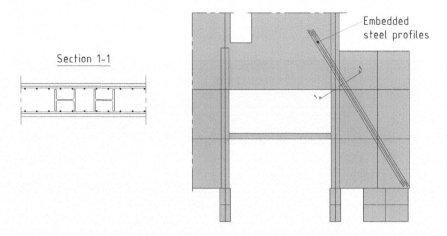

Section 1-1

Figure 1.5 Strengthening-inclined compression strut

structure locally hybrid was the chosen strengthening solution, as shown in Figure 1.5.

Figure 1.4 also shows lintels that connect the main core walls with external wing walls. These linking elements are heavily loaded in terms of shear and bending moment and are also typically present in the area where openings are required for mechanical, electrical and plumbing equipment (MEP). Specific embedded steel profiles, shown in Figure 1.6, were locally required to ensure that the building behaves laterally as appropriate. Significant cracks in these lintels, which would have existed if they had been made of reinforced concrete, would have led to an uncoupled-wall structural system, without significant transfer of

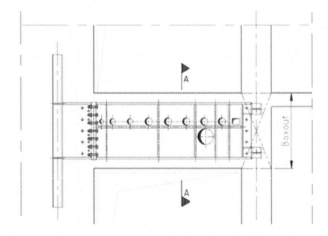

Figure 1.6 Coupling steel beams between core walls

Figure 1.7 Mohammed VI tower in Rabat, 250m height, 55 floors

shear via the lintels. This would have resulted in a smaller lateral resistance of the building and an unfavourable distribution of stresses at the foundation level.

Strong coupling beams were also needed for a high-rise tower in Morocco, as illustrated in Figure 1.7, because the limitation of lateral deformation under wind was one governing criterion in its design. The comfort of inhabitants was ensured by a tuned mass damper (TMD) located at the top, but the stiffness of the structure had to be considerable for the following reasons:

- Limitation of the inter-story drift to $h_s/300$ at each level to ensure the compatibility of the main structure's expected deformation with the acceptable façade deformation.
- Limitation of the general lateral deflection at the top to $H/450$ for a wind speed of 100-years return period in order to limit the higher bending moment of second-order effects expected at the bottom of the tower.

The lateral stiffness of the tower is therefore a key design parameter, and core walls have to work together, which requires strong coupling beams.

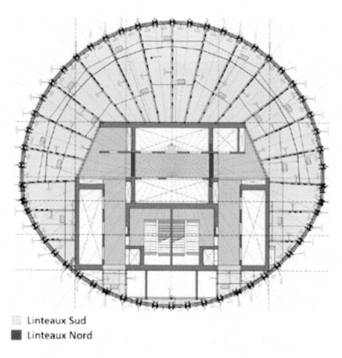

☐ Linteaux Sud
■ Linteaux Nord

Figure 1.8 Mohammed VI tower in Rabat – position of coupling beams

The plan view given in Figure 1.8 highlights the location of these beams, named "Linteaux Nord" and "Linteaux Sud" on the drawing. False ceilings are located just under those beams, and therefore no space below them is available for ducts required for MEP, so significant holes must exist in the coupling components. Classical concrete beams would not be acceptable, because significant cracking would occur, so it was decided that steel coupling beams be used, as illustrated in Figure 1.9. These are "Vierendeel" beams, which maximize space for MEP ducts. The action effects in each profile are a combination of moment, shear and axial force. Figure 1.10 shows the coupling components that have been mentioned.

Chapters 5 and 6 of this book give design guidance for the connection of the steel structures that are used to achieve a locally required optimum performance. The transition zone between the steel profile and the reinforced concrete section has to be adequately designed to ensure an effective transmission of axial forces, shear and bending moment carried by the interrupted steel element, illustrated in a simplified case in Figure 1.11.

The design of the new headquarters of Abu Dhabi's National Oil Company (ADNOC), United Arab Emirates, which is 342 metres in height and was built between 2010 and 2016, presented another technical challenge. A typical floor is shown in Figure 1.12. The vertical bearings elements are

Figure 1.9 Mohammed VI tower in Rabat – coupling beams connecting core walls

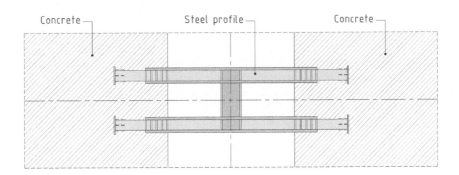

Figure 1.10 Details of coupling beams connecting core walls

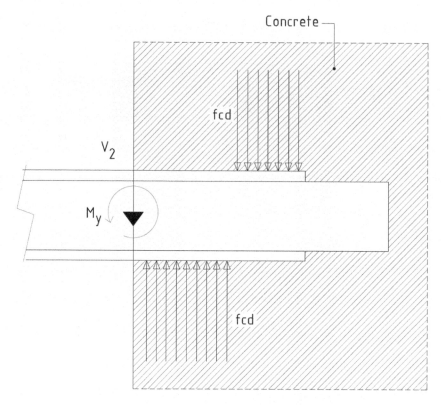

Figure 1.11 Transition zone between steel profile and concrete

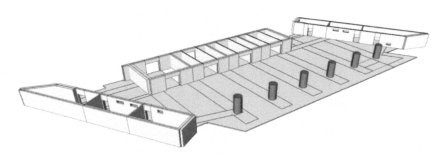

Figure 1.12 New headquarters of Abu Dhabi's National Oil Company – typical floor

three core walls and a row of columns. Due to architectural requirements, the size of these columns was limited, and typical reinforced concrete components were therefore not an option. For execution and quality reasons, a solution with concrete-filled tubes (CFT) was chosen. However, the design of specific beam-to-column moment connection details raised difficulties,

Figure 1.13 New headquarters of Abu Dhabi's National Oil Company – typical floor column connection

as there was little scientific knowledge and no specific design recommendations available at that time.

Chapter 6 of this book focuses on steel shear keys acting as connectors of reinforced concrete beams or flat slabs to steel columns. Figure 1.13 presents the solution realized in the 342-metre-high tower.

REFERENCES

SMARTCOCO. (2017) EUR 28914 EN, smart composite components - Concrete structures reinforced by steel profiles, European commission. Research programme of the research funds for coal and steel, TGS8 2016. RFSR-CT-2012-00031. ISBN 978-92-79-77016-6. https://doi.org/10.2777/587887; https://op.europa.eu/en/publication-detail/-/publication/aee33c6b-58b6-11e8-ab41-01aa75ed71a1

Plumier, A. (2016) Smartcoco project deliverable 2.2. Report on generic design approach (Internal report, available on request).

Chapter 2

Load introduction and force-transfer mechanisms at steel profile-concrete interface

Herve Degee and Rajarshi Das

CONTENTS

2.1 INTRODUCTION

The longitudinal shear force transfer at the connection between concrete and steel components is carried by four main mechanisms: a) chemical bond: the bond between the cement paste and the surface of the steel; b) mechanical bond: as the interface between the steel and the concrete is not

DOI: 10.1201/9781003149811-2

perfectly flat, there is a resistance to relative movements between the steel and the concrete; c) friction: assumed proportional to the normal force at the interface; d) mechanical interaction: due to ribs, shear studs or plate connectors. While chemical bonding is typically neglected in both design and analysis of composite structures, friction and especially mechanical actions (bonds and connectors) are very important (Salari, 1999).

Mechanical connectors are commonly used between a concrete slab and a steel section to provide the required composite action in flexure. They can also be used to distribute the large horizontal inertial forces in the slab to the main lateral load-resisting elements of the structure (Hawkins and Mitchell, 1984) or to enhance the connection between concrete and steel components in composite columns. Headed shear connectors are welded to the steel beam to provide different degrees of connection between the concrete slab and beam. Nevertheless, there are quite a few options when it comes to choosing a shear connector for the steel-concrete interface in a composite section such as embossments, ribs, plates and studs. The type of shear connection used between the steel and concrete elements has a significant effect on the behaviour of composite components. Rigid shear connectors, such as plates, usually develop full composite action. Flexible shear connectors, on the other hand, generally permit the development of only a partial composite action. So, consideration of interlayer slip between the components becomes of utmost necessity in the analysis procedure. The shear force transfer in composite components is experimentally measured using different techniques. The push-out test is the most common procedure because of its simplicity. Details of its basic layout and more recent enhancements can be found in Viest et al. (1997). Figure 2.1 shows qualitative shear resistance versus slip behaviour (Daniels and Crisinel, 1993), where two different behaviours can be noticed – "ductile" and "brittle". In both cases,

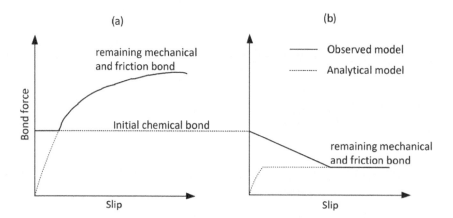

Figure 2.1 a, b Typical shear resistance versus slip behaviour: a) ductile response; b) brittle response. Copyright: SMARTCOCO

chemical bonding is indicated by bond resistance without accompanying slip. If we design such a composite section where the mechanical and frictional shear resistance is substantially lower than the initial chemical bonding shear resistance, we will only get a brittle behaviour (case b) and subsequently a brittle failure. This failure can be often observed in cases with decking with small or no embossments, and without shear studs. On the contrary, if we adopt embossed decks or simply add shear studs, the bond stresses associated with these mechanical devices prove to be much higher than the chemical bond stresses and as a result, a ductile response (case a) can be achieved (Plumier, 2016). In both cases, however, the initial chemical bond associated with no slip is neglected in design and analysis, as shown in Figure 2.1 by the dotted line.

2.1.1 Bond stress

For composite structures without mechanical connectors, most formulas assume that the bond stress distribution at the steel-concrete connection surface (or interface) is constant over the entire cross-section. That might necessarily be the case as observed from quite a few experiments done in the past. Roeder (1984) used the push-out test to study the bond transfer mechanism in embedded composite columns with no shear connectors. According to these analyses, the bond stress is primarily contributed by the flanges and the maximum bond stress is a function of the strength of the concrete. Wium and Lebet (1994) conducted similar tests on embedded columns. It was found that the average bond stress increases with the applied load (with basically no initial slip) until the peak point where chemical debonding occurs. After the peak, the bond stress decreases and then stabilizes on the value provided by frictional resistance (Figure 2.1b). The authors suggest using only the bond stress due to friction in the design and analysis of composite structures. This bond strength, which they call τ_{max}, depends on four major parameters: the thickness of the concrete cover, amount of hoop reinforcement, size of the steel section (depth of section) and concrete shrinkage. Wium and Lebet (1994) also distinguished between first and second loading. Their experimental tests on embedded columns showed that if a column is first loaded beyond the point of chemical debonding and is then unloaded and reloaded, even though the value of the maximum reachable bond stress τ_{max} does not change, the force-transfer mechanism and the stiffness of the shear stress-slip diagram change. The force transfer on the outer face of the flanges is different from the mechanisms between the flanges.

The bond stress capacity is commonly evaluated as the maximum average bond stress, which is the maximum load transferred between the steel and concrete, divided by the total surface area of the steel section embedded within the concrete. Many studies (Bryson and Mathey, 1962; Hawkins, 1973; Roeder, 1984; Roeder et al., 1999; Hamdan and Hunaiti, 1991;

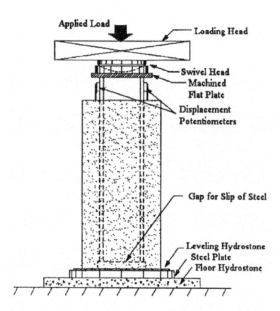

Figure 2.2 Typical push-out test (Roeder et al., 1999). Copyright: ASCE

Wium and Lebet, 1991; Wium and Lebet, 1992) have addressed the bond stress capacity of steel reinforced concrete (SRC) composite columns with push-out tests as illustrated in Figure 2.2. In the earliest push-out testing done by Bryson and Mathey (1962), the effect of the surface condition of the steel on the bond stress capacity was studied. The steel sections were freshly sandblasted, sandblasted and allowed to rust, or left with normal rust and mill scale. Sections that were sandblasted and allowed to rust developed larger maximum average bond stresses than those with mill scale. Once the steel core slipped relative to the concrete encasement, the post-slip bond stress was similar for all surface conditions. They postulated that the bond stress was primarily developed on the flange surfaces. A later study (Hawkins, 1973) examined the position of casting and the relative size of reinforcement. Specimens cast in the horizontal position had smaller bond capacity than those cast in the vertical position. This was attributed to the segregation of aggregate and accumulation of water under the lower flange of the horizontal steel section. The amount of confining reinforcement did not consistently affect the bond stress capacity prior to significant slip, but an increase in confining reinforcement increased the bond resistance after slip. The size of the steel section had no effect on the bond behaviour for specimens, provided that the ratio of the embedment length to the steel core depth was the same for all specimens.

Another study considered the distribution of bond stress over the component length (Roeder, 1984). Strain gauges were placed along the length of each specimen. The bond stress was found to vary exponentially under service loads, but the distribution was nearly uniform as the specimen approached its maximum capacity. Relative slip was observed and attributed to cracks developing in the concrete matrix along the concrete/steel interface. These results were qualitatively the same as those obtained on embedded plain steel rods under axial loading (Womersley, 1927) and torsion (Brown, 1966). It was postulated by Roeder that the specimen is permanently damaged by the onset of cracking and that load levels higher than the initial slip load should be avoided for cyclic loading. These tests suggested that the bond stress capacity increased with the concrete strength. However, later studies (Hamdan and Hunaiti, 1991; Wium and Lebet, 1992) disagreed with this observation.

Later push-out tests (Hamdan and Hunaiti, 1991) examined the effects of the concrete strength, surface condition, and tie reinforcement on the maximum average bond stress. The study suggested that concrete strength had no effect but that adding tie reinforcement to specimens with sandblasted steel surfaces increased the maximum average bond stress. The short-column test was employed in other studies, where the load is applied to the exposed steel at the top of the column but where the reaction is provided to both the concrete and steel at the base of the specimen. In a short-column test, as in a real column, the relative slip between the concrete and steel is limited to a portion of the column length, because the bottom part of the column has full composite behaviour provided by compatibility of the strains in the steel and concrete. The maximum average bond stress for short-column tests should however be based only on the load that is transferred from the steel to the concrete, rather than on the maximum load. Other researchers (Wium and Lebet, 1992) performed both short-column and push-out tests. They postulated that bond stress can be separated into two stages. The first stage occurs prior to complete slip and is governed by adhesion or chemical bonding between the cement paste and the steel. The second stage occurs after complete slip and is characterized as a purely frictional phenomenon. The tests showed that an increase in flange cover from 50 mm to 150 mm increased the force transfer after chemical debonding by 50%. Increasing the amount of hoop reinforcement also increased the bond stress capacity to some extent. Steel sections that were small relative to the concrete encasement had significantly larger maximum average bond stress, but concrete shrinkage was found to reduce the bond stress capacity slightly.

2.1.2 Scope of Chapter 2

While encased SRC construction has been used in numerous buildings, there is very little guidance for design in the United States. The American Concrete Institute (ACI) specifications have recognized the use of SRC columns since 1995 (ACI318, 1995; ACI Committee 318, 1995) but require that shear transfer between the steel and concrete be based entirely on direct bearing. No allowance is made for a natural bond between steel and concrete. The

American Institute of Steel Construction LRFD specifications permit the use of SRC construction in both beams and columns since 2005 (AISC360-5, 2005). The AISC provisions for columns require that all shear transfers be accomplished by direct bearing. Other countries have however established explicit design standards for SRC construction. European countries have established Eurocode 4 (2004), which permits the use of a natural bond of 0.3 MPa (see Table 6.6 of EN 1994-1-1) over the entire perimeter of the section for a length equal to the maximum dimension of the cross-section. The Japanese provisions permit the use of bond stress of $0.2f'c$ but not more than 0.45 MPa over the length of the component. Clearly, these specifications illustrate the significant differences from component specifications. This chapter discusses the various guidelines provided by Eurocode 4 on appropriate ways of load introduction and force-transfer mechanisms at the steel profile-concrete interface combined with certain recommendations based on recent experimental results as well as other international structural codes.

2.2 GENERAL APPROACH

A standard procedure to design concrete structures locally reinforced by steel profiles or concrete sections reinforced by several steel profiles is neither available in Eurocode 2 (2004) nor Eurocode 4 (2004). However, those standards offer certain guidelines for the load introduction and force-transfer mechanisms which can be directly applied at any steel profile-concrete interface irrespective of the structure type. A specific issue to be faced by the different possible solutions to transfer compression/tension forces from a steel profile to the surrounding concrete is to avoid creating local disturbances, like transverse cracking/splitting of concrete around the steel profile.

2.2.1 Experimental evidence from the SMARTCOCO research project

Ten tests were carried out considering six different configurations, including or not flexible and stiff connectors, different confinement schemes and different orientations of the steel profile with respect to the smallest dimension of the global composite specimen as listed below:

(1) *Configuration A*:	Steel profile with strong axis perpendicular to the longer wall face; connections with six flexible connectors on the total length of the steel-encased profile; transverse links at each connector.
(2) *Configuration B*:	Steel profile strong axis parallel to the longer wall face; connections with four stiff connectors ("transverse stiffener" type); transverse links and three different horizontal reinforcement schemes, denoted B1, B2 and B3.

(3) *Configuration C*: Steel profile weak axis perpendicular to the longer wall face; connections with six flexible connectors on the total length of the steel-encased profile; transverse links at each connector.

(4) *Configuration D*: Steel profile weak axis perpendicular to the longer wall face; connections with four stiff connectors ("transverse stiffener" type); transverse links and two different reinforcement schemes, denoted D1 and D2.

(5) *Configuration E*: Steel profile weak axis perpendicular to the longer wall face; connections with two flexible and two stiff connectors on the total length of the steel-encased profile; transverse links at each connector.

(6) *Configuration F/G*: Steel profile strong axis perpendicular to the longer wall face; respectively with rust and paint on profile (no mechanical connectors).

Direct push-out tests were performed on all the specimens. Figure 2.3 shows a schematic diagram as well as an example of the real test set-up and load application for a clearer understanding. Figure 2.4 presents some photographs of the prepared test specimens. The connectors and the vertical and horizontal reinforcements of all the configurations were designed based on specific guidelines available in Eurocode 2 and 4. A brief overview of the different configurations and their results are described throughout this chapter in relevant sections.

From all the experiments, some general observations can be made, for example, the Eurocode 4 design rules offer a safe estimation for the specimens with headed studs, i.e. configurations A and C, independently of the orientation of the steel profile (web-oriented parallel or perpendicular to

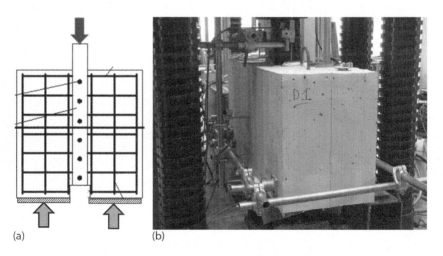

(a)   (b)

Figure 2.3 a, b Experimental set-up for push-out tests: (a) schematic diagram (b) real-life set-up

Figure 2.4 Prepared test specimens before and after concrete casting

the wall faces) inside the concrete wall. The ultimate resistance obtained from the experiments was found to be 40 to 50% greater than the design resistance estimated based on Eurocode 4. The analytical as well as experimental values are precisely presented in Table 2.1 for all the specimens. A ductile behaviour was also noticed. Based on such observations, it appears that the Eurocode 4 rules can be safely applied to connections using headed shear stud connectors. Furthermore, the Eurocode 4 rules were also seen to provide appropriate results regarding the bond resistance for encased profiles. Certain limitations were however noticed. Therefore, new modified rules are recommended based on the experimental results to ensure a better general design of the composite sections. The existing general rules available in Eurocode 4 are discussed below along with the additional recommendations necessary to design concrete sections reinforced by several steel profiles.

2.2.2 General provisions for composite sections

The available provisions generally used to estimate the resistance of composite cross-sections assume that no significant slip occurs at the interface between the concrete and structural steel components. However, the design provisions applicable for conventional composite solutions provide certain principles (e.g. EN 1994-1-2005, Clause 6.7.4.1(1)) for limiting slip in the critical regions of load introduction for (a) internal forces and moments applied from components connected to the ends and (b) for loads applied within the length to be distributed between the steel and concrete components, i.e. basically where axial load and/or bending moments are applied to the column, considering the shear resistance at the interface between steel and concrete. For the structural configurations considered herein, it is additionally deemed necessary to provide a clearly defined load path not involving any significant amount of slip at the steel-concrete interface, which would invalidate the preliminary assumptions made in design. In

Table 2.1 Resistance to longitudinal shear at steel profile-concrete interface – analytical and experimental*

Spec.	Connectors	Transverse reinforcement	Web orient.	V_{Rd} bond (kN)	V_{Rd} connect (kN)	V_{Rd} total (kN)	$V_{R,Exp}$ (kN)	$\dfrac{V_{R,Exp}}{V_{Rd\,(Total)}}$
A	Six studs on flanges	One link/stud	//	232	434	666	1050	1.57
B1	Four plates	No	//	0	678	678	600	0.88
B2	Four plates	Six hoops	//	232	678	910	920	1.01
B3	Four plates	Six links	//	232	678	910	740	0.81
C	Six studs on flanges	One link/stud	⊤	232	434	666	910	1.37
D1	Four plates	No	⊤	232	678	910	1000	1.10
D2	Four plates	Six links + six hoops	⊤	232	678	910	1000	1.10
E	Two plates + two studs on flanges	Six links + six hoops	⊤	232	484	716	700	$0.98 \approx 1.00$
F	0-rust	No	//	501	0	501	1480	2.95
G	0-paint	No	//	0-(501)*	0	0-(501)*	580	∞ (1.16)

* Where V_{Rd} is the relevant design resistance analytically calculated based on Eurocode 4 and $V_{R,Exp}$ is the resistance values obtained from the experiments at a 0.5 mm slip.

general scenarios, most shear connectors do not reach their design shear strength until a 1 mm slip takes place at the interface. Such a slip is rather insignificant for a resistance model based on plastic behaviour and rectangular stress blocks. However, once the load path is actually a "long path", it implies a greater global slip and therefore requires further control such as, for instance, that the assumed load path should stay within the introduction length limit given in code provisions, like for instance in Clause 6.7.4.2(2) of Eurocode 4.

After defining a clear load path (as per the Eurocode 4 provisions), it is possible to estimate shear stresses at the interface between steel and concrete. When composite columns and/or compression components are subjected to significant transverse shear either by local transverse loads or by end moments or both, they give rise to longitudinal shears as well. Then, suitable provisions should be made to transfer these longitudinal shear stress at the steel-concrete interface.

It should be noted that the shear stresses often exceed the design shear strength (Clause 6.7.4.3) in the regions of load introduction, and therefore an additional shear connection is required (Clause 6.7.4.2(1)), as stated in the earlier paragraphs. It can also occur at definite locations where the shear strength τ_{Rd} (from table 6.6 of Eurocode 4) is very low or when the composite section has a high degree of double curvature bending.

In general, for axially loaded columns and compression components, the existing version of the Eurocode 4 suggests that there is no need to consider longitudinal shear outside the areas of load introduction, which means that, even for columns checked for buckling, we do not need to consider a deformed situation involving end moments and the corresponding longitudinal shear at the steel-concrete interface. Although this is correct for composite sections with a central symmetrical steel section under pure compression, due to the absence of any transverse and longitudinal shear, it is not entirely true for a compression component submitted to end moments. The aforementioned experiments done on the composite sections with several embedded steel profiles highlighted a few critical consequences. As a result, it can be concluded that longitudinal shear is indeed essential to be considered inside and outside the areas of load introduction for sufficient safety.

Furthermore, the longitudinal shear developed due to bending at the steel-concrete interface in the composite columns or walls reinforced with multiple embedded steel profiles showcased significant importance through the experimental observations and therefore should be evaluated based on either an elastic or a plastic model. Sufficient shear resistance should then be provided following the design guidelines discussed in Chapters 3 to 6 of this book.

Additionally, longitudinal shear should be considered inside the area of load introduction in order to distribute the external applied loads or the reaction forces at connections. This can be achieved according to the suggestions of Section 2.3.

2.3 LOAD INTRODUCTION IN HYBRID COMPONENTS

The primary concern that arises while introducing or applying the loads to a hybrid component or a composite section is to achieve a successful transfer of shear forces from one material surface to the other, for example from an embedded steel profile's surface to the concrete elements surrounding it. In today's construction industry, there exist quite a few elements or "shear connectors" to achieve such a force-transfer mechanism such as headed shear studs, perfobond ribs, T-rib connectors and welded plate connectors. For the last six decades, however, the headed shear studs have been the most popular among all these options.

2.3.1 Shear stud connectors – force transfer

One of the most illustrative models to explain the load transfer through stud connectors in solid slab applications was given by Lungershausen (1988) where four different components which contribute to the total capacity of the connector were defined (Figure 2.5). Initially, the majority of the longitudinal shear force is transferred at the base of the stud into the surrounding concrete (A) where a significant amount of it reacts directly at the weld collar. The multi-axial high-bearing stresses in the concrete eventually lead to local crushing failure of the concrete at the bottom of the stud and to a redistribution of the shear forces in areas higher up the shank of the stud (B). Since the top of the stud is embedded in undamaged concrete and cannot deform while the base of the connector is free to move laterally, bending and tensile stresses are induced into the shank of the stud (C).

To balance these tensile stresses, compressive forces develop in the concrete under the head of the stud and are thought to activate additional frictional forces (D) at the steel-concrete interface. Eventually, the shear connection fails when the shank of the stud experiences a combined shear-tension failure right above the weld collar. The model is suitable to describe the main factors influencing stud behaviour. The area of the stud shank

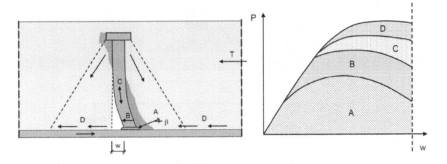

Figure 2.5 Load transfer of a headed stud connector in a solid slab in accordance with Lungershausen (1988). Copyright: Ruhr-Universität

and its ultimate tensile strength obviously influence the shear strength of the stud directly.

The longitudinal shear flow in composite steel and concrete beams is transferred across the steel-flange-concrete-slab interface by the mechanical action of the shear connectors, which simply act as steel dowels that are embedded in a concrete medium. The ability of the shear connection to transfer longitudinal shear forces, therefore, depends on the strength of the dowel action in longitudinally uncracked slabs and, also, on the resistance of the concrete slab against longitudinal cracking induced by the high concentration of load imposed by the dowel action (Oehlers and Park, 1992). The mechanism (Johnson and Oehlers, 1981) by which the dowel action transfers the longitudinal shear forces is illustrated in Figure 2.6 for the case of a stud shear connector.

Shear forces F_{sh} and flexural forces M_{sh} are induced in the shank of the stud. These forces are in equilibrium with an eccentric normal force across the stud-shank-concrete-slab interface acting at a distance b_f from the steel-flange-concrete-slab interface. The shank of the stud is, therefore, subjected to shear and flexural stresses, and the concrete zone immediately in front of the stud is subjected to high compressive stresses. The magnitude of these stresses depends not only on the shear force but on the position b_f of the resultant normal force, which is a function of the stiffness of the concrete relative to that of the steel. For example, if the stiffness of the concrete E_c tended to infinity, then the eccentricity b_f would tend to zero, and, similarly, if E_c tended to zero, then b_f, would tend to half the height of the stud, i.e. the normal stress across the shank-concrete interface would be uniformly distributed. The dowel strength of the shear connection depends therefore on the strength and stiffness of the stud material and also on the compressive

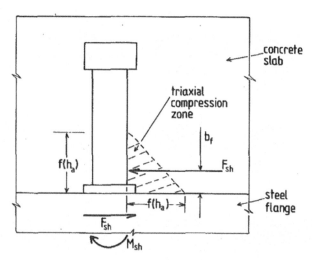

Figure 2.6 Dowel action (Oehlers and Park, 1992). Copyright: ASCE

strength and stiffness of the concrete in the zone directly in front of the stud. Several other research studies have also provided different analytical and numerical models which could successfully predict the load-transfer behaviour of the shear studs and have therefore confirmed their efficiency as shear connectors. As a result, standard design procedures have been documented in the structural codes all around the world, and Eurocode 4 is no exception. Eurocode 4 provides specific clauses regarding the necessity of shear connectors when loads are introduced to hybrid components. In this section, some of these clauses are considered, which apply very well to the composite sections with one or multiple steel profiles embedded inside a concrete component.

2.3.2 Provisions for loaded composite sections

Eurocode 4 already recommends some straightforward guidance regarding the general requirements of all types of composite sections. For instance, whenever the design shear strength τ_{Rd} (as defined in Clause 6.7.4.3 of Eurocode 4) is exceeded at the steel-concrete interface in the load introduction areas or in component transition zones, i.e. areas with changing cross-sections, shear connectors are absolutely necessary. The shear forces should also be determined from the change of sectional forces of the steel or reinforced concrete section within the introduction length. In certain cases, for example, when the axial load is applied only to the steel component, the force transferred to the concrete can be estimated from the relative axial loads in the two materials. An accurate calculation is quite difficult where the cross-section does not govern the design of the column. On the other hand, when the axial force is applied to both materials or only the concrete, the force percentage resisted by the concrete gradually decreases due to creep and shrinkage. In these situations, elastic and fully plastic models would provide a safer result. However, in absence of a more accurate method, we should not consider an introduction length larger than $2d$ or $L/3$, where d is the minimum transverse dimension of the column and L is the column length.

For composite columns and compression components, no shear connection needs to be provided for load introduction by endplates if the full interface between the concrete section and endplate is permanently in compression, taking account of creep and shrinkage. Otherwise, the load introduction should be verified according to the Eurocode 4 guidelines (Clause 6.7.4.2(1) to be specific). For concrete-filled tubes of circular cross-section, the effect caused by the confinement may be considered if the conditions given in Clause 6.7.3.2(6) of Eurocode 4 are satisfied using the values η_a and η_c for λ equal to zero.

Where stud connectors are attached to the web of a fully or partially concrete-encased steel I-section or a similar section, frictional forces develop due to the prevention of lateral expansion of the concrete by the adjacent

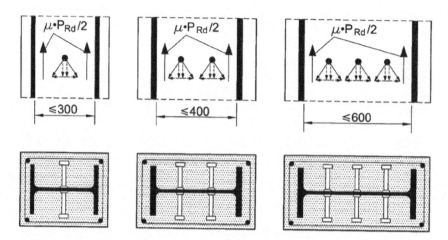

Figure 2.7 Additional frictional forces in composite columns by use of headed studs (Eurocode 4 – Figure 6.21). Copyright: European Commission

steel flanges. Eurocode 4 recommends considering these frictional forces and adding the frictional resistance to the calculated resistance of the shear connectors. In such a case, the additional resistance can be assumed to be $\mu\,P_{Rd}$ on each flange and each horizontal row of studs, as shown in Figure 2.7, where μ is the relevant coefficient of friction. For steel sections without painting, μ may be considered as 0.5. P_{Rd} is the resistance of a single stud in accordance with Clause 6.6.3.1 of Eurocode 4. In absence of better information from tests, the clear distance between the flanges should however stay within the values given in Figure 2.7.

If the cross-section is partially loaded (as in the example shown in Figure 2.8), the loads may be distributed with a ratio of 1:2.5 over the thickness of the end plate. The concrete stresses should then be limited in the area of the effective load introduction, for concrete-filled hollow sections in accordance with Clause 6.7.4.2(6) of Eurocode 4 and for all other types of cross-sections in accordance with EN 1992-1 Section 6.7.

The push-out tests on configurations A and C, presented in the previous section and illustrated in Figure 2.9, offered concrete validations to the aforementioned clauses and also provided some additional information regarding the transverse reinforcements required for such composite sections.

It was observed that, for H profiles oriented with the web parallel to the longer wall face, transverse reinforcement by hoops around the profiles was necessary to equilibrate the compression struts forces. Although for H profiles oriented with the web perpendicular to the wall face, transverse reinforcement was not required for the internal profiles, transverse reinforcement by hoops around the profiles was required for the profiles situated in the confining reinforcement within edge regions of the cross-section

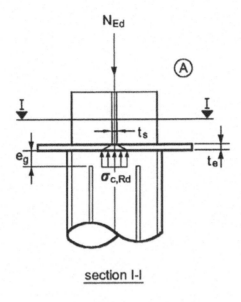

section I-I

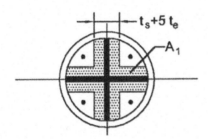

Figure 2.8 Partially loaded circular concrete-filled hollow section (Eurocode 4 – Figure 6.22). Copyright: European Commission

(termed "boundary elements" in Eurocode 8 (2004)). Standard provisions were therefore documented based on these test results and are explained in Chapter 4 with more details.

In the case of load introduction through only the steel section or the concrete section, for fully encased steel sections, the transverse reinforcement should be designed for the longitudinal shear that results from the transmission of normal force (N_{c1} in Figure 2.10) from the parts of concrete directly connected by shear connectors into the parts of the concrete without direct shear connection (see Figure 2.10, section A-A; the hatched area outside the flanges should be considered as not directly connected). The design and arrangement of transverse reinforcement should be based on a truss model assuming an angle of 45° between concrete compression struts and the component axis.

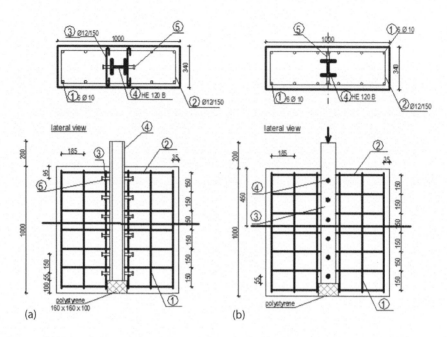

Figure 2.9 a, b Cross-sections for (a) Configuration A and (b) Configurations C

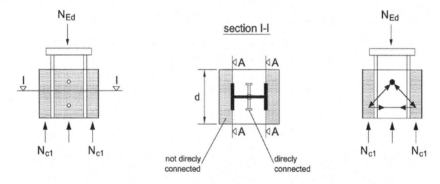

Figure 2.10 Directly and not directly connected concrete areas for the design of transverse reinforcement (Eurocode 4 – Figure 6.23). Copyright: European Commission

2.4 LONGITUDINAL SHEAR IN HYBRID COMPONENTS OUTSIDE OF THE AREA OF LOAD INTRODUCTION

This section provides an overview of the necessary application rules corresponding to the development of longitudinal shear in composite columns and compression components when they are subjected to significant transverse shear. As can be seen from the values in table 6.6 of

Eurocode 4, the design shear strengths τ_{Rd} are quite small compared to the tensile strength of concrete. Therefore, the separation at the steel-concrete interface level is actually governed by the frictional characteristics rather than the bond itself. For instance, if supposing that the concrete has expanded laterally around a partially encased I-section, although it will create pressure on the flanges, it would not do so on the web and that is why the design value stated for the webs of a partially encased section is set to τ_{Rd} equal zero. Similarly, concrete expansion would apply the highest pressure when it is inside a hollow steel tube. Therefore, the highest shear strengths are provided for cases having concrete inside steel tubes.

2.4.1 Provisions to deal with longitudinal shear

The available provisions suggest that, outside the area of load introduction, longitudinal shear caused by transverse loads and/or end moments should always be verified at the interface between the concrete and the steel. Shear connectors should be provided, based on the distribution of the design value of longitudinal shear, where this exceeds the design shear strength V_{Rd}.

It is also recommended that, as long as no more accurate method is available, an elastic analysis considering the long-term effects and cracking of concrete may be used to determine the longitudinal shear at the interface. Corresponding to this requirement, Chapter 4 presents a method developed for the design of composite columns/walls with several encased steel profiles. The experimental results obtained from configurations F and G (as shown in Figure 2.11) provided detailed information regarding the pure bonding characteristics between the embedded steel profile and the surrounding concrete material. The steel profile was placed inside the concrete with its strong axis perpendicular to the longer wall face, and no mechanical connectors were used. The steel profile surface was artificially rusted in Configuration F and smoothly painted in Configuration G to examine each of their frictional properties with respect to table 6.6 of Eurocode 4.

As shown by the curves in Figure 2.11, the post-peak bond resistance offered by Configuration F was approximately $\tau_{R,\,exp}$ = 1480 kN / 0,686 m^2 = 2,15 N/mm^2, which is almost three times greater than the value obtained when calculated according to the design recommendations of Eurocode 4 (i.e. $\beta c * \tau_{Rd}$ = 2.5 x 0.3 = 0.75 N/mm^2). Even if we consider a partial safety factor of 1.5 on the experimental results, the estimated resistance is: $\tau_{R,exp}$ /1.5 = 2.15/1.5 = 1.43 N/mm^2, still two times greater than the Eurocode 4 design value. Such observations actually indicated that the basic τ_{Rd} value (= 0.3) available in Eurocode 4 might cause great underestimation for steel profiles encased in the concrete wall. On the other hand, the bond resistance

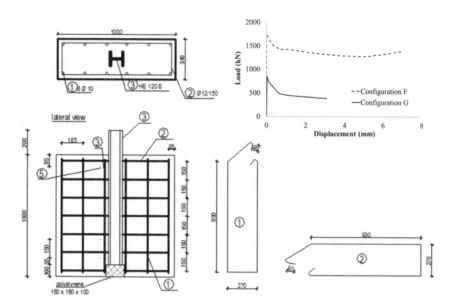

Figure 2.11 Configuration F and G – cross-sections and push-out behaviour

offered by the painted specimen (Configuration G) was approximately $\tau_{R, exp}$ = 580 kN / 0,686 m² = 0,84 N/mm². Although Eurocode 4 indicates that nothing reliable can be stated if there is paint or loose rust or grease on the surface of the embedded steel profile and does not provide any design value (τ_{Rd}) for such cases, the values obtained from the experiments were not at all insignificant. It can even be seen that the bond resistance using anti-rust paint was actually greater than the basic Eurocode 4 design value $\beta c * \tau_{Rd}$ = 2.5 x 0.3 = 0.75 N/mm². Therefore, further design values are suggested in addition to the design values already available in table 6.6 of Eurocode 4 as shown in Table 2.2.

Moreover, some recent push-out tests carried out at Hasselt University with profiles without any specific surface finishing have led to an average bond resistance of 1.26 N/mm² with a residual friction resistance of 0.71 N/mm², intermediate between the two limit cases of the SMARTCOCO tests F and G.

As a conclusion of the previous observations, and provided that the surface of the steel section in contact with the concrete is unpainted and free from oil, grease and loose scale or rust, the values given in Table 2.2 are suggested to calculate V_{Rd}; the first five lines in this table are similar to those of table 6.6 of Eurocode 4.

The value of τ_{Rd} given in Table 2.2 for completely concrete-encased steel sections applies to sections with a minimum concrete cover of 35 mm and transverse and longitudinal reinforcement in accordance with Chapter 4 of

Table 2.2 Design shear strength τ_{Rd}

Type of cross-section	τRd (N/mm²)
Completely concrete-encased steel sections	0.30
Concrete-filled circular hollow sections	0.55
Concrete-filled rectangular hollow sections	0.40
Flanges of partially encased sections	0.20
Webs of partially encased sections	0.00
Completely encased steel sections with paint	0.20
Completely or partially concrete-encased steel sections with oil, grease, loose scale or rust	0.00

this book. For greater concrete cover and adequate reinforcement, higher values of τ_{Rd} may be used. Unless verified by tests, for completely encased sections the increased value $\beta_c * \tau_{Rd}$ may be used, with β_c given by:

$$\beta_c = 1 + 0.02c_z \left(1 - \frac{c_{z,min}}{c_z}\right) \le 2.5 \qquad (2.1)$$

where,

c_z is the nominal value of concrete cover in mm, see Figure 2.12;

$c_{z, min}$ = 35 mm is the minimum concrete cover. The upper bound simply means that increasing the concrete cover c_z above 115 mm is not improving the situation further.

For partially embedded steel profiles under weak-axis bending due to lateral loading or end moments, there remains quite often a possibility that the encasement would separate from the web. For such cases, it is recommended to provide shear connectors for partially encased I-sections, unless

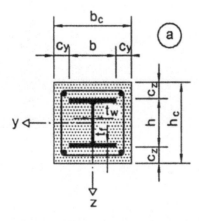

Figure 2.12 Definition of symbols (Eurocode 4 – Figure 6.17a). Copyright: European Commission

it is verified otherwise. If we consider that the resistance to transverse shear is provided by some additional elements in the structure rather than the structural steel alone, then the required transverse reinforcement for the shear force $V_{c,Ed}$ should be welded to the web of the steel section, according to Eurocode 4 (Clause 6.7.3.2(4)), or it should pass through the web of the steel section.

2.5 RESISTANCE TO LONGITUDINAL SHEAR AT THE STEEL-CONCRETE INTERFACE

The shear force at the connection zone between the steel and concrete components is fundamentally transferred by four main mechanisms: a) chemical bond: the bond between the cement paste and the surface of the steel; b) mechanical bond: the interface between steel and concrete is not perfectly flat and there is a resistance to relative movements between steel and concrete; c) friction: assumed proportional to the normal force at the interface; and d) mechanical connectors: providing the required interaction via ribs, shear stud connectors or plates. Chemical bonding is a bond resistance without accompanying slip. It is sensitive to the applied loading scenario and is generally neglected in design and analysis because, if it acts alone, a brittle behaviour may be observed. Mechanical bond and friction, on the other hand, are shear resistances which exist even if there is slippage, and so they are considered in the design and analysis of composite structures. Eurocode 4 provides the necessary provisions to consider mechanical bond, giving in its table 6.6 values of bond resistance τ_{Rd} which, in part, are those of Table 2.2 of this book.

The resistance to shear stresses at the steel-concrete interface can be calculated as a combination of mechanical bond, friction and connectors. The friction resistance can result from the application of a force external to the structural element and normal to its surface or from compression struts internal to the hybrid structural element. The friction resistance should be calculated with a friction coefficient μ equal to 0.5, for steel without painting.

The design shear strength can further be magnified by a factor β greater than 1,0 if the concrete cover considered is greater than 40 mm. Nevertheless, the reader can observe that the values of the design shear strength τ_{Rd} only apply if the concrete cover and transverse and longitudinal reinforcement defined in the code are present, which indirectly guarantees the existence of a force normal to the surface, necessary to mobilize friction. Furthermore, it also requires some surface roughness, which in fact means some mechanical interaction.

Mechanical interaction, due to embossments, ribs or shear stud connectors, is a shear resistance which exists with both stiff connectors (little or no slippage) and flexible connectors (higher slippage). For the last 60 years, headed shear stud (flexible connector) has been by far the most common type of shear connector used in the design of composite components in steel frame construction. It had also been one of the fundamental components of any composite component in which it had been used. So, a thorough understanding of the performance characteristics of the shear studs is arguably essential to achieve an efficient and reliable design of composite components. Studies on shear stud connectors started in the 1950s when the critical load of the stud was presented on the basis of push-out tests. As mentioned in the introductions of this book as well as this chapter, push-out tests are commonly used to determine the capacity of the shear connection and load-slip behaviour of such shear connectors. A calculation model of stud shear bearing capacity was first presented by Ollgaard et al. (1971) at Lehigh University which actually became the basis of several design methods followed in today's codes. From 1970 until today, numerous researchers and engineers have further tried to obtain more accurate relationships of stud shear bearing capacity and also to analyze the influence of geometrical and mechanical parameters on stud shear bearing capacity. Thanks to them, some important parameters have been identified which have an influence on the behaviour of the shear stud connectors such as shank diameter, height and tensile strength of studs, compressive strength and elastic modulus of concrete and reinforcement detailing. A detailed overview regarding each individual parameter's influence is however not discussed in this book but can be found in a recently published research study (SMARTCOCO, 2018).

Due to limited information, Eurocode 4 fails to provide any specific physical insight into the design shear strength, let it be bond or friction or mechanical interaction. It only covers the friction aspects with some basic mechanical interaction due to the minimal embossment constituted by the roughness of a lightly rusted steel surface. Available up-to-date research studies however indicate that friction resistance and mechanical transfer, which both allow slippage, could be combined to resist longitudinal shear at the steel-concrete interface. The push-out tests performed within SMARTCOCO on configurations combining shear studs and/or plate connectors, like A, C and E (shown in Figures 2.9 and 2.13) and presented in Section 2.2 can be used as examples to understand and further clarify some of the withstanding peculiarities. While Section 2.7 shows how the strut-and-tie models can be applied to the shear studs' design for Configuration A and C, Figure 2.13b presents the strut-and-tie model for Configuration E. A detailed overview of designing plate connectors is given in Section 2.6. The most fundamental conclusions were pretty straightforward. The provisions of Eurocode 4 can be safely applied to the configurations with only headed stud shear connectors, independently of the orientation of the steel profile (web-oriented parallel or perpendicular to the wall faces). Adding the shear resistance μ · τ_{Rd} at the steel-concrete interface to the resistance coming from the shear stud connectors provided an analytically calculated resistance still significantly lower than the experimental strength, which means

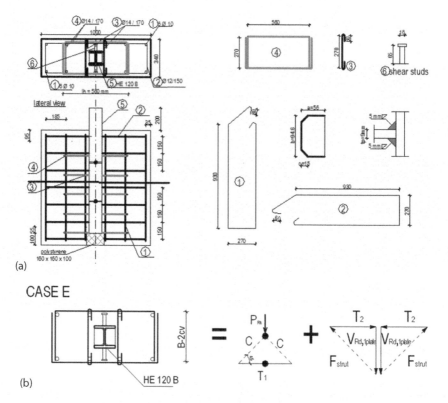

(a)

CASE E

(b) HE 120 B

Figure 2.13 a, b Configuration E: (a) cross-sections and (b) strut-and-tie model used to design the shear studs

that the "general indication" mentioned just above, can be considered as a safe and practical approach. Therefore, the contribution to shear resistance provided by the bond and the individual stud connectors can be added to achieve the necessary longitudinal shear resistance. For Configuration E, adding the resistance of studs and plate connectors provided a correct estimate of the real strength and it was therefore duly concluded that the Eurocode 4 provisions could be used in an acceptable manner along with the design method discussed in the later section to combine the studs and plate connectors.

Based on the previous statements and on additional experimental evidence available in Degee et al. (2016), the recommended design procedure can be summarized as follows:

(1) The resistance provided by headed shear connectors or studs should be calculated according to Clause 6.6.3.1 of Eurocode 4, and such connectors should be provided where the design value of longitudinal

shear exceeds the design shear resistance provided by bond and friction.

(2) The resistance provided by plates welded to an encased steel profile should be calculated as indicated in Section 2.6.

(3) With headed studs, if the distance from the wall or column surface to the connector is less than 300 mm, measures should be taken to prevent longitudinal splitting of the concrete caused by the shear connectors.

(4) With welded plate connectors, measures should be taken to prevent spalling of the concrete if the compression strut developed at the connector is directly facing a wall face.

(5) Stirrups or links should be placed at each connector. They should be designed to resist a tension force equal to the shear capacity of the connector.

2.6 WELDED PLATES SHEAR CONNECTORS

In the previous sections, two types of mechanical connectors have been considered, i.e. headed shear studs and welded plate connectors. The shear resistance due to headed stud connectors has been studied extensively, as explained in Section 2.3, and requires no new attention, the shear resistance brought by stiffeners between the flanges of the H section (stiff connectors) or to ribs welded on the web of H sections requires a specific approach. Steel plates simply fall under this category as one of such stiff connectors, and in order to design them, we first need to have a good understanding of their behaviour while used in composite and hybrid components.

A lot of literature can be found regarding steel plates used in a concrete retrofitting context, leading to composite solutions as plates glued at the external surface of existing concrete elements. The failure behaviour of a plate in such a composite section is governed by the plate's integrity with the original concrete component through adhesive bonding, which basically acts as the shear connector between the plate and the original RC component. Therefore, bond strength becomes the most important issue in the design of effective retrofitting solutions using externally bonded plates. Oehlers (2001) studied and subsequently categorized several types of failure mechanisms for these bonded plates. His research focussed on the bond strength and the critical bond length, which is defined as the length of the externally bonded plate. He specifically found that there is no further increase in the axial load-carrying capacity of the plate-adhesive-concrete interface beyond the bond length. In the same year, Ali et al. (2001) studied similar composite systems and stated that a detailed knowledge of the bond behaviour of the plate-adhesive-concrete interface is absolutely necessary to understand the failure/debonding mechanism of the plates. However, studies by Chen and Teng (2001) unfortunately highlighted the lack of a unique design procedure for the determination of the concrete-adhesive-plate interfacial bond strength and critical bond length for metallic bonded plates. From a general perspective though, certain observations and failure mechanisms were clearly noticed from these previous research investigations. The bond failure largely occurs due to crack propagation near or along the adhesive-concrete interface, parallel to the length of the plate. The crack initiates in the vicinity of the most stressed end and propagates further towards the free end of the plate. The debonding in such a case is defined as shear debonding failure. An experimental investigation was conducted in 1985 by Rabbat and Russell (1985) to determine the bond strength and the coefficient of static friction between a rolled steel plate and cast-in-place concrete or grout (Figure 2.14).

Fifteen push-off tests were performed on concrete specimens having different bond strengths between 0.17 and 0.61 MPa. The bond strength was found to be negligible in the case of the grout specimens. The average effective coefficient of static friction for concrete cast on steel plate and grout cast below steel plate was obtained as 0.65 for a wet interface and 0.57 for a dry interface. The normal compressive stresses in both cases ranged between 0.14 and 0.69 MPa. Several tests have also been conducted on both circular and square concrete-filled tubes to study the bond between the concrete and steel tubes (Shakir-Khalil, 1993a, 1993b; Lu and Kennedy, 1994; Hajjar et al., 1998), which could indirectly contribute to the steel-concrete shear bond stress-slip behaviour. However, they are not further discussed herein in detail in order to keep our focus solely on the plate used as shear connectors, for which much fewer references are available.

Although in today's industry, shear studs are already accepted as "the" shear connectors, specific cases might arise where "stiff shear connectors" such as plates provide a more efficient solution. Thanks to some aforementioned studies as well as to some individual provisions and design

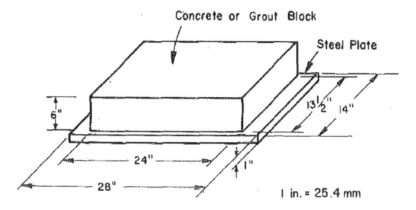

Figure 2.14 Test specimen (Rabbat and Russell, 1985). Copyright: ASCE

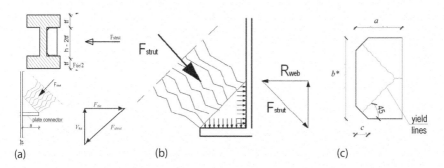

Figure 2.15 a, b, c (a) Strut-and-tie model to determine plate connector strength, (b) a normal pressure *p* is applied by the compression struts to the web and lower flange and (c) internal bearing plate yield line pattern (three fixed sides condition)

approaches found in the structural codes, it could be possible to try to formulate a design procedure for the welded plate connectors, pre-design test specimens and subsequently validate the formulated design procedure based on the experimental results. However, repeating this procedure every time needed would not be an option for industrial designers. A specified analytical procedure could provide the utmost efficiency. The following paragraph offers a formulated design procedure to design welded plate connectors for composite sections with embedded steel profiles.

Stiffeners in the form of plates welded on a wall of an encased steel profile, like for instance between the flanges of a H section as shown in Figure 2.15a, can have a direct bearing on concrete compression struts and provide resistance to longitudinal shear. As such, they can be designed according to the strut-and-tie method defined in Eurocode 2 and recalled in Section 2.7 of this book.

Let's state that the plate has the following geometrical properties:

Width of the plate: $a = \dfrac{b_f - t_w}{2}$

Length of the plate: $b^* = h - 2t_f$

Width of the clipped corners: c

and $A_{plate} = a \times b^* - c^2$

Then, assuming that the struts are formed at an angle $\theta = 45°$, the strut width and individual strut resistance F_{Rd} can be calculated as: $\dfrac{a}{\cos\theta} = \dfrac{a}{\sqrt{2}/2}$

$$F_{Rd} = a\sqrt{2}\sigma_{c,Rd,max}b^*$$

where $\sigma_{c,Rd,max} = 0.6v'f_{cd}$ and $v' = 1 - \dfrac{f_{ck}}{250}$

Therefore, the resistance to longitudinal shear V_{Rd} can finally be obtained as:

$$V_{Rd} = F_{strut}cos\theta = F_{strut} / \sqrt{2} = ab^* \times \sigma_{c,Rd,max} \tag{2.2}$$

When using multiple bearing plates, the minimum spacing between the plates should be at least two times the plate width a, in order to allow the full development of the concrete compression struts inclined at 45°. On the other hand, the maximum proposed spacing can be stated as $6a$.

If a longitudinal shear, $V_{l,Ed}$, is applied to the composite section, the minimum required number of plates n_{plates} over a certain length l is found as:

$$n_{plates} = \frac{V_{l,Ed}}{V_{Rd}} \tag{2.3}$$

The design of the plate connectors, which are horizontal supports to the compression struts, is made considering the normal pressure applied by the struts to the plate as shown in Figure 2.15b. Therefore, the plate should be checked for shear and bending under that pressure. If V_{Ed} represents the applied shear per unit length of the plate connector, then the pressure p applied to the plate can be calculated as $p = V_{Ed}/A$, where A is the rectangular area of the plate equal to ab^*.

Two approaches can be used to define the necessary plate thickness t_p, one referring to an elastic design and another to a plastic design with yield lines forming in the plate. In the elastic analysis approach, bending moments in the supporting plates should be calculated using elastic equations of bending moments m_{Ed} per unit length in plates for definite support conditions, three fixed sides and one free side, according, for instance, to Table 2.3.

An elastic stress σ can then be calculated as $\sigma = m_{Ed} / W_{el}$

where W_{el} is the elastic bending modulus per unit length of plate $W_{el} = t_p^2 / 6$

The elastic resistance may be preferred to the plastic one ($W_{pl} = t_p^2 / 4$) in order to limit deformations of the supporting plates at the ultimate limit state of the compression struts and to provide enough overstrength to those plates.

Both fillet and butt welds can be used for such connectors. Nevertheless, all welds should be designed to have an overstrength with respect to the connected plates, to avoid any issue due to failure in the welded connection. The weld throat a_w of fillet welds should therefore comply with the condition $2a_w > t_p$. Furthermore, hoops or cross-ties should be placed in the concrete section to equilibrate the strut compression force.

Next to the elastic approach, a fully plastic approach based on yield lines can also be followed to define the necessary plate thickness t_p. This approach can be made using for instance Roark's formulas for stress and strain, which are based on yield lines in the plate connector (Figure 2.15c)

Table 2.3 Elastic bending moments m_{Ed} per unit length in plates with three fixed sides and one free side (l = smallest between a and b)

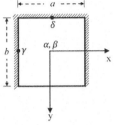

| | Centre of the plate | | Middle of the fixed edge | |
| | $M_x = \alpha q l^2$ | $M_y = \beta q l^2$ | $M_x = \gamma q l^2$ | $M_y = \delta q l^2$ |
b/a	α	β	γ	δ
0.5	0.0206	0.0554	−0.0783	−0.1140
0.6	0.0245	0.0481	−0.0773	−0.1020
0.7	0.0268	0.0409	−0.0749	−0.0907
0.8	0.0277	0.0335	−0.0708	−0.0778
0.9	0.0274	0.0271	−0.0657	−0.0658
1.0	0.0261	0.0213	−0.0600	−0.0547
1.1	0.0294	0.0204	−0.0659	−0.0566
1.2	0.0323	0.0192	−0.0705	−0.0573
1.3	0.0346	0.0179	−0.0743	−0.0574
1.4	0.0364	0.0166	−0.0770	−0.0576
1.5	0.0378	0.0154	−0.0788	−0.0569
1.6	0.0390	0.0143	−0.0803	−0.0568
1.7	0.0398	0.0133	−0.0815	−0.0567
1.8	0.0405	0.0125	−0.0825	−0.0567
1.9	0.0410	0.0118	−0.0831	−0.0566
2.0	0.0414	0.0110	−0.0833	−0.0566
∞	0.0417	0.0083	−0.0833	−0.0566

and are referred to in AISC341-10 (AISC341-10, 2010). Assuming that $b \geq 2a$ and $t_p < t_f$, the required bearing plate thickness t_p should satisfy:

$$t_p \geq \frac{2.8a}{a + 0.9b^*} \sqrt{\frac{2a^2 p (3b - 2a)}{3\phi f_{ay} (6a + b^*)}} \qquad (2.4)$$

where f_{ay} is the plate yield stress and $\phi = 0.9$. The welds can remain designed in a similar manner as in the elastic approach; hoops or cross-ties should again be placed to equilibrate the strut compression force.

A design procedure still does not hold its value if not properly validated. To that purpose, five specimens have been tested during the SMARTCOCO

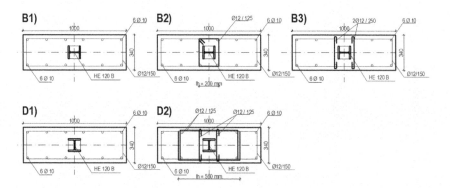

Figure 2.16 Cross-sections of the different specimens for Configuration B and D

project, three for Configuration B (web of the profile parallel to the long dimension of the concrete element) and two for Configuration D (web of the profile perpendicular to the long dimension of the concrete element), as introduced in Section 2.2 and illustrated in Figure 2.16. The "stiff" or welded bearing plate connectors were designed based on the aforementioned procedure and applied to both these configurations. Different reinforcement and confinement schemes were used along with a different orientation of the embedded steel profile in order to get a clear understanding of any critical parameters.

Configuration B1 and D1 were constructed without the ties, normally needed to equilibrate the compression struts developing from the plate. Hoops or cross-ties were provided around the steel profile and across compression struts in configurations B2 and D2. The relative load-displacement behaviour between the concrete and the free edge of the steel profile obtained from Configuration B and D are shown in Figures 2.17(a) and 2.17(b) respectively.

Conclusions may be drawn from those experimental results. It can first be noticed from Figure 2.17(a) and Table 2.1 that, when the compression struts are perpendicular to the wall faces (web of profile parallel to longer wall face – Configuration B), the closed hoops used in specimen B2 are absolutely necessary to achieve the calculated resistance. The experimental resistance obtained in this case is almost perfectly the one predicted by the analytical calculations. On the other hand, confinement by links connecting the mesh, as used in specimen B3, proves to be less effective, as the experimental resistance is 740 kN instead of a calculated resistance of 920 kN. Specimen B1 performs even worse, evidencing the need for tying the compression struts.

On the contrary, Figure 2.17(b) shows that when compression struts develop parallel to the wall faces (web of profile perpendicular to longer wall face – Configuration D), the same resistance is obtained with or

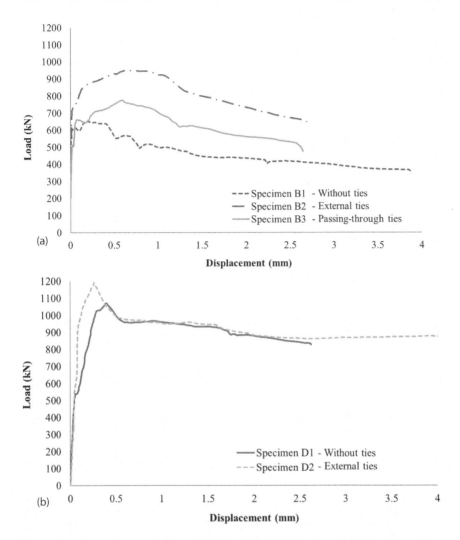

Figure 2.17 a, b Relative displacement between concrete and free edge of the steel profile in (a) specimens B1, B2 and B3 and (b) specimens D1 and D2

without transverse reinforcements. Specimens D1 and D2, both provide an equal experimental yield resistance of 1000 kN. This happens because, in these cases, the compression struts have more volume to spread into and therefore can find steady anchorage in the steel meshes parallel to the faces of the wall. Comparing Table 2.1 and Figure 2.17(b), it can also be noticed that the proposed design method for this type of connector provides a safe sided estimate of the resistance values, i.e. 900 kN compared to the experimental resistance of 1000 kN. The transverse reinforcements could only slightly increase the stiffness and the peak resistance but did not influence the experimental yield resistance.

2.7 ABOUT STRUT-AND-TIE

Actual standards, such as Eurocode 2 (CEN, 2004), allow the use of the strut-and-tie method to design concrete components, for "continuity" regions as well as "discontinuity" regions. Continuity regions, also called "B" regions, are the parts of the components where the Bernoulli hypothesis is applicable. Discontinuity regions, called "D" regions, are the other parts. Following the Saint-Venant principle, D regions are located around any cause of localized effects: punctual load, support, change in the geometry – they extend in general over a length equal to the maximum of the height and the width of the component section (see Figure 2.18).

The "strut-and-tie" method is a convenient design tool that can be used to handle the transition zones between composite and reinforced concrete components, as will be illustrated in Section 2.8 and in Chapter 5. However, some caution is necessary.

In theory, strut-and-tie models should be developed based on the stress fields obtained by the theory of elasticity. Compressive principal stresses are aggregated in struts, while tensile stresses are condensed in ties. The arrangement of the steel reinforcement is then defined in accordance with the position and direction of the ties. This implies a reorientation of forces to consider that rebars (and ties) will be often placed in orthogonal directions that do not correspond exactly to tensile principal stresses. This modification supposes thus a redistribution of stresses, which is allowed considering the ductility of the steel and the relative ductility of the concrete, at least in compression. However, this reorientation must be limited, around 15 ° according to Schlaich et al. (1987).

Often practitioners don't take the time to develop a full linear finite element model for every detailing design. They conceive strut-and-tie models based on classical examples found in the codes or in the literature (Schlaich et al., 1987; fib, 2011a, 2011b; CEN, 2004) or using simplified methods as the load path method, based on stress trajectories, as illustrated in Figure 2.19. By doing so, they refer to their experience and their intuitive knowledge of structural behaviour.

It is often said that, if the strut-and-tie model satisfies the equilibrium and if the resistance of all the struts, nodes and ties is provided, the obtained design is safe based on the lower-bound theorem of plasticity. This theorem states that a stress field that satisfies equilibrium and does not violate yield criteria at any point provides a lower-bound estimate of capacity of elastic-perfectly plastic materials.

It is worth remembering here that, as concrete is not an elastic-perfectly plastic material, the lower-bound theorem does not apply and that the

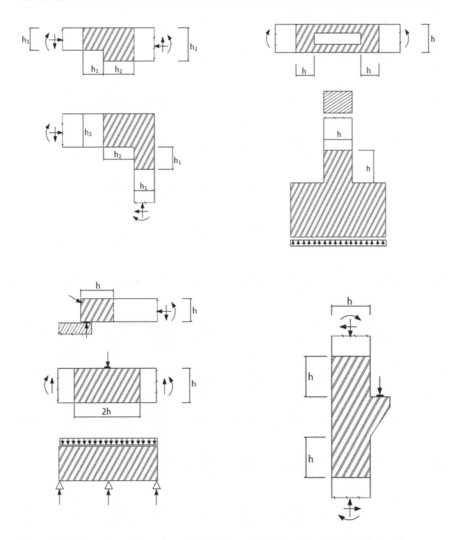

Figure 2.18 Examples of D regions (Schlaich et al., 1987). Copyright: PCI Journal

application of simplified methods for the development of structural models must be made carefully. Section 2.8 will give an illustration of a case where a redistribution that, at first sight, seemed reasonable, led to an unsafe design.

2.8 INDIRECT SUPPORT OF A STEEL BEAM

In this section, a work that handled the particular point of secondary steel beams crossing a primary concrete beam is presented (Marie and Somja,

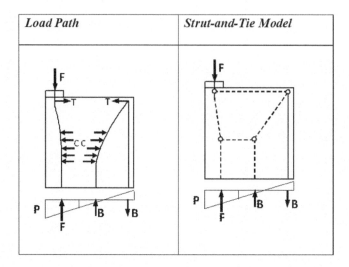

Load Path	*Strut-and-Tie Model*

Figure 2.19 Illustration of the load path method (Schlaich et al., 1987). Copyright: PCI Journal

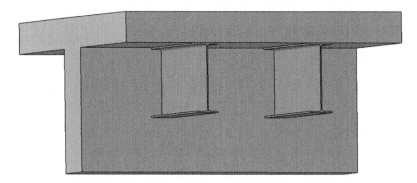

Figure 2.20 Indirect support of steel beams supporting a slab

2018), see Figure 2.20. The title of this section refers to the title of the corresponding paragraph in Eurocode 2, named "indirect support", applicable when the primary and secondary beams of beams are both made of concrete.

A hybrid zone commonly exists when, in usual concrete construction, steel beams are used to attain large spans. In order to make on-site building as simple as possible, the transfer of the forces from the steel to the concrete is made by simply embedding the steel secondary beam in the concrete primary beam, so that no specific skills in steel construction are needed to set up the beam.

In the following, the experimental tests achieved on that line in the SMARTCOCO project are first described. Then, a specific strut-and-tie model is deduced from elastic stress trajectories and validated against experimental results. It is finally extended by a parametrical study in order to propose guidance for the design.

2.8.1 Experimental evidence from the SMARTCOCO project

The specimens consist of a concrete beam crossed in its middle by a steel profile, see Figure 2.21. The load is applied to the steel profile by two jacks, one at each end of the steel profile. Moreover, a horizontal reaction is applied at each end of the specimen in order to increase the capacity of the concrete beam by arch effect.

Five tests were carried out, with two different layouts of the secondary steel beam:

- Two specimens with a HE340M profile, named CS-H.
- Three specimens with a HE340M profile with median horizontal stiffeners, named CS-H+R.

The width of the steel flanges had been reduced in the embedded part of the steel beam, in order to limit the resistance in the investigated zone. The main dimensions, dimensions of the steel section and rebar arrangement are given respectively in Figures 2.22, 2.23 and 2.24.

The load-displacement curves are presented in Figure 2.25. All specimens, with or without transverse stiffeners, behave the same way. The differences observed for CS-H-1 and CS-H+R-3 are merely due to variations in the horizontal reaction effectively applied to the specimen. Indeed, for specimen CS-H+R-3, a larger horizontal initial force has been imposed before starting the vertical loading, see Figure 2.26.

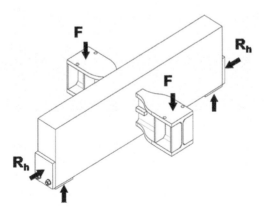

Figure 2.21 Specimen and loading

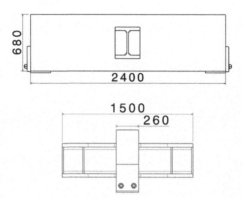

Figure 2.22 Main dimensions of the specimen

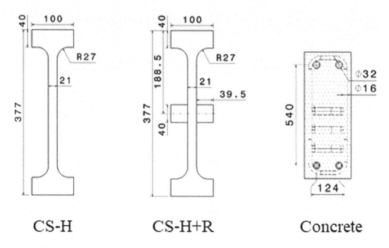

CS-H   CS-H+R   Concrete

Figure 2.23 Dimensions of the steel and concrete sections

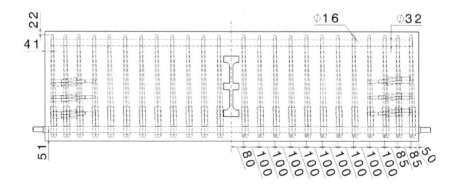

Figure 2.24 Rebar arrangement

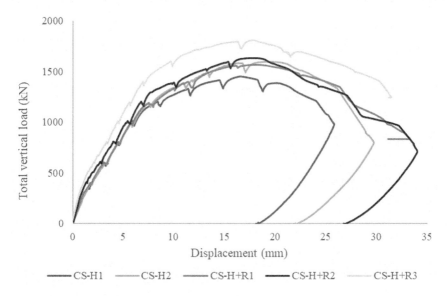

Figure 2.25 Vertical force-displacement curves

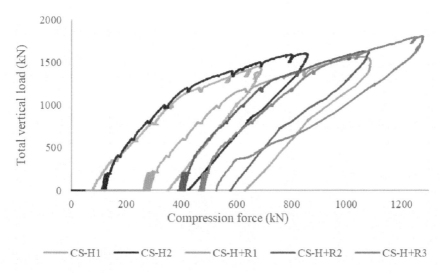

Figure 2.26 Horizontal force-vertical force curves

The same behaviour has been observed for all the specimens. First, bending is dominant: vertical cracks are detected beneath the steel profile. Next, shear cracks with an orientation of about 45°, starting from the steel profile and going to the bottom side of the specimen, develop. For a load of about 1200 kN, vertical cracks beneath the steel profile open largely, showing

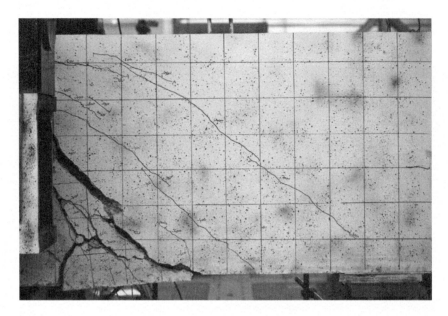

Figure 2.27 Specimen CSH-R+3 after collapse and shape of the punching cone

that the longitudinal bars yield. This is confirmed by the evolution of the strains measured by strain gauges. Anyway, this first yielding does not lead to the full collapse of the specimen, as the horizontal support allows the development of an arch effect in the concrete beam. A maximum load ranging from 1400 to 1800 kN is attained, after which a punching cone develops; the test is stopped when the first stirrup breaks. Figure 2.27 shows specimen CS-H+R3 after collapse.

The evolution of the stresses in the stirrups of the specimen CHS-R+3 is shown in Figure 2.28. As in other specimens, it can be observed that only the first stirrups near the steel profile have yielded.

2.8.2 Interpretation of the results

The initial design of the specimens was based on a strut-and-tie model drawn with two major hypotheses, see Figure 2.29:

- An inclination of 45° with respect to the vertical was considered for the diagonal struts.
- The lower-bound theorem of plasticity was supposed applicable, and a uniform distribution of the forces in the different stirrups was considered, neglecting a possible effect of the variable lengths of the diagonal struts.

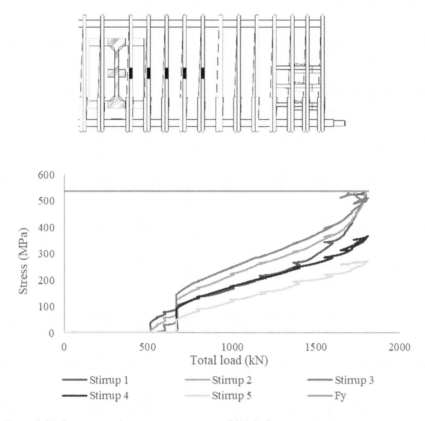

Figure 2.28 Stresses in the stirrups: specimen CSH-R+3

This model lead to an overestimation of the collapse load because these two hypotheses were invalidated by the experimental evidence:

- The shape of the punching cone, Figure 2.27, shows that the angle of diffusion is less than 45°.
- The distribution of the forces in the stirrups at collapse is not uniform, as shown in Figure 2.28.

It is thus mandatory to reconsider the definition of the strut-and-tie model, on a more academic basis. As explained in Section 2.6., strut-and-tie models must be based on stress distributions obtained from linear elastic analysis. Such an analysis was performed for the test specimen CS-H+R using the finite element software FINELG (de Ville et al., 2016). A total vertical force of 1668 kN, the mean of the maximum experimental loads, is applied together with a horizontal force of 1000 kN, the mean of the horizontal loads. Stress trajectories found are presented in Figure 2.30. The position of the stirrups and longitudinal bars is marked by black lines numbered 1 to 5.

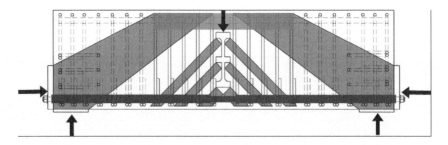

Figure 2.29 Initial model invalidated by the test

This model shows clearly that struts are not parallel and that their inclination varies with the level of the load applied to the beam.

A new local strut-and-tie model was thus defined, see Figure 2.31. As the sum of the forces in the first four stirrups balance the load applied by the jacks, the fifth stirrup is not considered active. Struts are drawn from every horizontal steel-concrete contact surface to every base of the stirrups, within the diffusion area set in evidence by the elastic analysis.

The evolution of the forces in the stirrups is compared to the experimental results in Figure 2.32, and it shows that the correlation is good.

The fact that the fifth stirrup is considered as not active can be justified by two reasons:

- First of all, the real strut, which can be visualized by the cracking, is more straight than elastic stress trajectories and corresponds to the secant line drawn as a dotted line at 51° shown in Figure 2.30. As

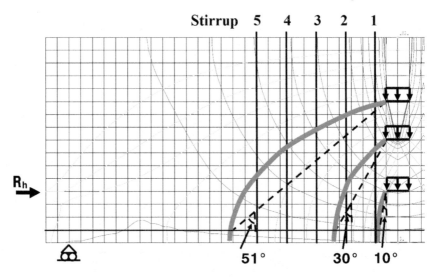

Figure 2.30 Elastic stress trajectories for specimen CSH-R+3 and inclinations θ of compression struts on the vertical corresponding to these trajectories.

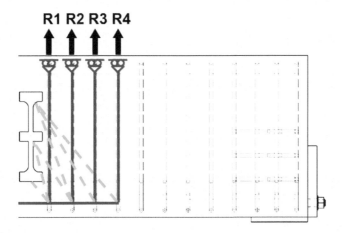

Figure 2.31 New strut-and-tie model

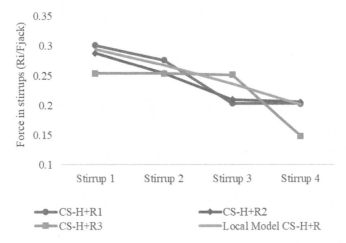

Figure 2.32 Comparison of forces in the stirrups

a consequence, the anchoring length of the stirrup in the strut is not large enough to develop yielding in the stirrup;

- Simple analytical reasoning shows that the force in the stirrups varies with $\cos^3\theta$, where θ is the inclination of the strut on the vertical, see Figure 2.33 and Eqs (2.5), (2.6):

$$N_{strut} = EA\frac{\Delta L}{L} = \frac{EA_{strut}\delta\,\cos(\theta)}{d/\cos(\theta)} \qquad (2.5)$$

$$N_{stirrup} = N_{strut}\cos(\theta) = \frac{EA_{strut}}{d}\delta\,\cos^3(\theta) \qquad (2.6)$$

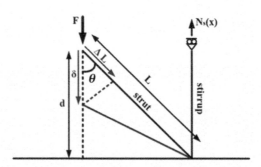

Figure 2.33 Effect of the inclination θ of the strut on the force in the stirrup

where A_{strut} is the section of the strut, δ is the vertical displacement of the point of application of the load, θ is the inclination of the strut on the vertical and d is the distance from the applied load F to the lower longitudinal rebar.

As a consequence, the forces in the stirrups in the elastic model decrease rapidly with the distance to the mid-section of the primary beam and the force in the fifth stirrup is negligible.

In conclusion, the analysis of the results shows that the strut-and-tie design model must be carefully drawn from elastic stress trajectories and that the inclination of the struts bounding the loads to the suspending stirrups plays a key role in the distribution of the forces, by limiting the contribution of the stirrups which are the most distant from the mid-section of the primary beam.

2.8.3 General model for indirect support

Before defining a general model, a point must still be investigated. Elastic stress trajectories in Figure 2.30 show that the level of application of the load modifies substantially the angle of diffusion θ. Given the evidence from the previous section, this angle of diffusion can be assimilated to the inclination of the compression struts, justifying the same notation θ for both quantities. This θ is very small when the load is applied near the lower flange of the concrete beam, and even when the load is applied at mid-height, θ is still limited to 30°. This restricts the number of stirrups that can support the steel profile.

A parametrical study has been made in order to get general information on the evolution of the angle of diffusion with the level of application of the load. Several beams with different lengths have been computed by elastic linear analysis, and the angle of diffusion has been deduced from the stress trajectories. As can be seen in Figure 2.34, the angle θ can be as low as 10° when the load is applied in the lower half of the specimen.

However, recommending an angle of diffusion θ of 10 ° when the load is applied in the lower part of the section would be too restrictive, and not in

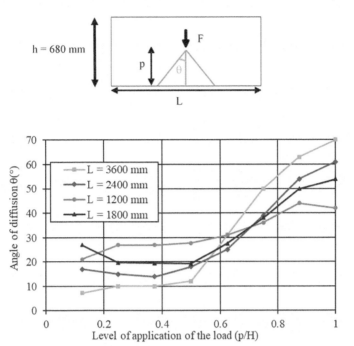

Figure 2.34 Determination of the total force in the stirrups

line with actual standards. It should be remembered that the elastic stress trajectories must not be considered exact, and a variation of 15° can be adopted following Schlaich et al. (1987). As a consequence, it is proposed to adopt an angle θ of 18°, as proposed in Clause 9.2.5, "indirect support", of Eurocode 2 (CEN, 2004) for the case of a secondary concrete beam. And, if the lower flange of the steel beam is above 0.7 h_c, it is proposed to open the angle θ up to 30°.

This angle may still seem quite small, but it was also chosen to maintain a good efficiency of all the stirrups within the diffusion zone, knowing that the force in the stirrups located within the angle of diffusion cannot be considered as uniform, as has been shown by experimental evidence.

The ultimate load is then defined by the yielding of the first stirrup, and the ultimate load is defined from the total yielding force of the stirrups penalized by a reduction factor ρ. ρ is computed supposing that the stirrups can be replaced by a vertical tie distributed uniformly along the length of the beam, see Figure 2.35. The force can be computed as the integral of the forces in the tie:

$$\frac{F}{2} = \int_{x_i}^{x_f} N_s(x)\,dx = \int_{x_i}^{x_f} a_s f_y \frac{\cos^3\theta(x)}{\cos^3\theta_i(x)}\,dx \qquad (2.7)$$

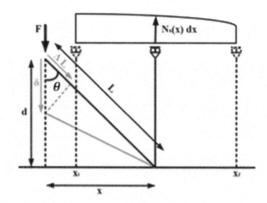

Figure 2.35 Determination of the total force in the stirrups

That gives

$$F = \rho A_s f_y \tag{2.8}$$

with

$$\rho = \frac{1}{x_f - x_i} \int_{x_i}^{x_f} \frac{\cos^3 \theta(x)}{\cos^3 \theta_i(x)} dx \tag{2.9}$$

with A_s is the total section area of the stirrups within the angle of diffusion, a_s is the distributed section, x_i is the abscissa of the beginning of the zone of diffusion at the level of the longitudinal rebar, x_f is the abscissa at the end of the angle of diffusion and θ_i is the angle of the strut corresponding to x_i.

The evolution of ρ with the maximum angle of diffusion calculated considering $x_i = 0$ is shown in Figure 2.36. As can be seen, the reduction factor for an angle of diffusion of 30° is around 0.9.

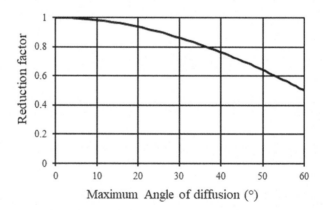

Figure 2.36 Evolution of the reduction factor ρ of the stirrup resistance with the angle of diffusion θ.

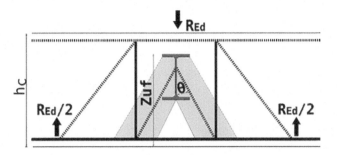

Figure 2.37 Strut-and-tie model for the indirect support of a steel beam inside a concrete beam

2.8.4 Proposal of simplified design guidelines

A simplified design procedure is proposed, based on the general model presented in 2.8.4 and on the strut-and-tie scheme in Figures 2.37 and 2.38:

1. The inclination of the struts is limited to $\tan(\theta) \leq 0.3$ (equivalent to $\theta < 17°$).
2. If the height of the upper flange z_{uf} is greater than $0.7\, h_c$, where h_c is the height of the concrete beam, the inclination θ of the struts may be increased up to $\tan(\theta) \leq 0.6$ (equivalent to $\theta < 30°$).
3. In this latter case, the resistance of the ties F_{td} must be reduced by a factor 0.9 to consider the non-uniform distribution of the stresses in the different stirrups.
4. The inclination of the struts depends on the general configuration of the steel and concrete beams. The minimum values given in 2. and 3. may be increased on the basis of the stress trajectories obtained with linear elastic analysis. In the model, each stirrup must then be linked by an independent strut to the steel profile in order to consider the non-uniform distribution of the tension force among the different stirrups.

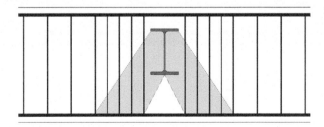

Figure 2.38 Arrangement of reinforcement for the indirect support of a steel beam inside a concrete beam

2.9 DESIGN EXAMPLES

This section presents the application of the previously described design approaches to ensure the appropriate longitudinal shear transfer capacity between an embedded steel profile and the surrounding concrete. Four options are considered:

- Using shear studs.
- Using welded plate connectors.
- Using a combination of shear studs and welded plate connectors.
- Relying only on the bond resistance between steel and concrete.

Problem statement and general data

Design of a 5 metre long composite column with an adequate longitudinal shear resistance between the embedded steel profile and the surrounding concrete.

The general assumptions for the preliminary design calculations are listed below:

Materials:

- Concrete (C30/35).
  - Characteristic strength f_{ck} = 30 MPa.
  - Mean value of the elastic modulus E_{cm} = 32000 MPa.
- Embedded steel profile HE 200B.
  - Steel profile grade S460.
- Reinforcement bars Φ18 S500.
- Material partial factors are stated equal to 1.0 for the sake of an easy presentation. For a practical design, the values of $_{\gamma v}$, $_{\gamma M0}$, $_{\gamma c}$ and $_{\gamma s}$ recommended by the Eurocodes and their national annexes should be applied.

Concrete column dimensions:

The minimum width B of the column can be calculated as:

$$B = 2 \cdot c_v + 2 \cdot \phi_l + h_{HE\,200B} + 2 \cdot h_{sc}$$

$$= 2 \cdot 35\,mm + 2 \cdot 18\,mm + 200\,mm + 2 \cdot 65\,mm = 436\,mm$$

B = 450 mm
  with:

- Concrete cover c_v = 35 mm.
- Diameter of the longitudinal reinforcement Φ_l = 18 mm.
- Shear stud height h_{sc} = 50 mm.
- Width of the concrete column D = 450 mm.

- Depth of the concrete column B = 450 mm.
- Height of the concrete column H = 5000 mm.

The system is designed to transfer a total load of 2500 kN from the steel profile to the surrounding concrete, corresponding to 70% of the axial plastic resistance of the profile.

Option 1 – Load transfer by shear studs
This solution corresponds to the most conventional way to activate the longitudinal shear transfer mechanism at the steel-concrete interface, i.e. by using shear studs as mechanical connectors.

The characteristics of the shear studs are as:

- Diameter of the shear studs d = 16 mm.
- Stud height h_{sc} = 50 mm, so that $3d$ = 48 mm $\leq h_{sc}$.
- Longitudinal spacing s_c where $5d$ = 80 mm $\leq s_c \leq$ min $(6h_{sc}; 800\text{mm})$ = 390 mm.
- Maximum stud tensile strength f_u = 450 MPa.

The strength of the connection is then obtained from Eurocode 4 Part 1-1 §6.6.3.1. (1) as:

$$P_{Rk} = min\left(\frac{0.8 \cdot f_u \cdot \pi \cdot \dfrac{d^2}{4}}{\gamma_V}, \frac{0.29 \cdot \alpha \cdot d^2 \cdot \sqrt{f_{ck} \cdot E_{cm}}}{\gamma_V}\right) = min\left(72.38\text{kN}, 60\text{kN}\right) = 60\text{kN}$$

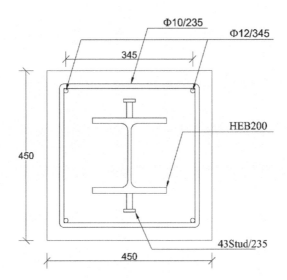

Figure 2.39 Solution I – using shear studs

where:

$d = 16$ mm, $h_{sc}=50$ mm, $f_u = 450$ MPa and $\gamma_v = 1.00$ (see remark in the data regarding γ_v)

$$\alpha = \begin{cases} 0.2\left(\dfrac{h_{sc}}{d}+1\right) \text{ for } 3 \le \dfrac{h_{sc}}{d} \le 4 \\ \qquad 1 \text{ for } \dfrac{h_{sc}}{d} > 4 \end{cases} = 0.825;$$

The necessary number of shear studs is $\dfrac{N_{max}}{P_{Rk}} = \dfrac{2500\,kN}{60\,kN} = 42\,pcs$

The strength capacity of $n_{studs} = 21$ shear studs/side is equal to $N_{Rd} = n_{studs} \cdot P_{Rk} = 2520$ kN.

The distance between two shear studs is equal to $s_c = 235$mm.

The basic design requirements for longitudinal and transverse reinforcements can be obtained from Eurocode 2 Part 1-1 Clause 9.6.

- the area of longitudinal reinforcement should be between

$A_{s.vmin} = 0.002 \cdot A_c = 405\,mm^2$ (with min 4Φ12) and

$A_{s.vmax} = 0.04 \cdot A_c = 8100\,mm^2$.

- the distance between two longitudinal bars should not exceed three times the column thickness or 400 mm; $S_v = 450 - 2 \cdot 35 - 2 \cdot 14 - 12 = 340\,mm$.
- the area of transverse reinforcement should not be less than
$A_{s.hmin} = 0.001 \cdot A_c = 202.5\,mm^2 = $ min 2Φ12.
- the distance between two transverse hoops should not be more than 400 mm, $S_h < 400$ mm.

Additionally, it is also required to verify that the transverse reinforcements are able to work as ties to anchor the compression struts in the concrete induced by the load transfer from the shear studs. This is carried out according to the strut-and-tie model given in Figure 2.40(a), assuming an angle of the struts on the vertical of 45°.

The tying force per shear stud is $T = \dfrac{P_{Rk}}{2} = 30\,kN$. The corresponding amount of transverse reinforcements is then obtained as $A_{sh} = \dfrac{T}{f_{sd}} = 0.60\,cm^2$.

Consequently, 1Φ10 hoop every 230 mm is a possible solution.

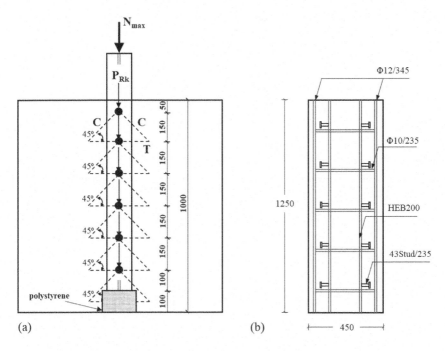

Figure 2.40 Solution I – (a) strut-and-tie model, (b) side view

$$\frac{H}{s_c} = \frac{5000\,mm}{230\,mm} \approx 21 \text{ hoops should be provided as indicated in Figure 2.40b,}$$

which gives a total area of $A_{sh} = 2 \cdot A_{21\phi10} = 32.97\,cm^2$.

Option 2 – Load transfer by stiffener-like welded plate connectors
In this solution, steel plates are used as mechanical connectors between the embedded steel profile and the concrete column.

The plate connectors are welded to the web of the embedded steel profile, similar to classical stiffeners. The following geometrical characteristics (Figure 2.41a) are obtained for the plate connectors used in this solution:

- Width of the plate $a = \dfrac{b_f - t_w}{2} = 95.5\,mm$.

- Length of the plate $b = h - 2 * t_f = 170\,mm$.

- Width of the clipped corners $c = 19\,mm$.

- Area $A_{plate} = a * b - c^2 = 159.11\,cm^2$.

The force transfer associated with the plate connectors is determined considering the activation of the strut-and-tie mechanism shown in Figure 2.42. It is considered that the struts are formed with an angle, $\theta = 45°$.

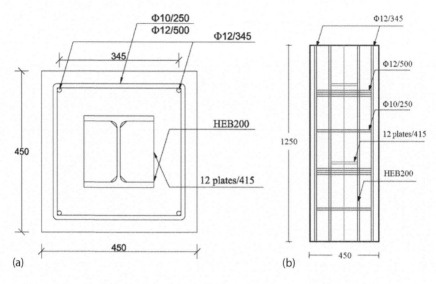

Figure 2.41 a, b Solution 2 – using plate stiffeners (a) top view and (b) side view

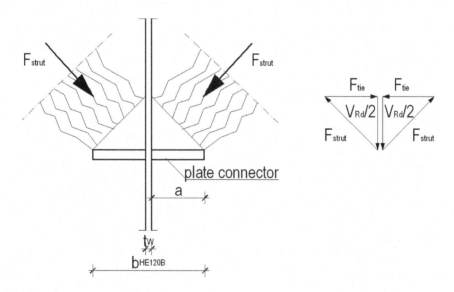

Figure 2.42 Strut-and-tie model to verify the plate connectors

The strut width can be calculated as: $\dfrac{a}{\cos\theta} = \dfrac{a}{\dfrac{\sqrt{2}}{2}} = 135.06\,\text{mm}$ with a

corresponding strut resistance $F_{Rd} = \dfrac{a}{\sqrt{2}} \cdot \sigma_{Rd,max} \cdot b = 257.2\,\text{kN}$

Where $\sigma_{Rd,max} = 0.6 \cdot \upsilon' \cdot f_{cd} = 15.84\,MPa$, f_{cd} =   30   MPa   and

$\upsilon' = 1 - \dfrac{f_{ck}}{250} = 0.88$

The transfer force activated by a single plate is $V_{Rd,1plate} = F_{Rd} \cdot \cos\theta = 181.9\,\text{kN}$

The number of plates required can be calculated as $n_{plates} = \dfrac{N_{max}}{V_{Rd,1plate}} = 13.75$

The total strength capacity of n_{plates} = 14 plate connectors, 7 plates on each side, is equal to:

$N_{Rd} = n_{plates} {}^{*} V_{Rd,\,1plate} = 2547\,\text{kN}.$

The resistance of the connector itself is based on theoretical solutions for plates supported on three sides. The required bearing plate thickness is determined on the basis of the equation used in Plumier et al. (2013). The yield lines form at 45° from the corner, meaning that the yield line is shorter than $2a\sqrt{2}$, as shown in Figure 2.43. The actual yield line length is $b\sqrt{2}+a-b/2 = 250.92$ mm.

The bearing pressure on a plate is given by $w_u = \dfrac{V_{Rd,1plate}}{A_{plate}} = 11.43\,MPa$

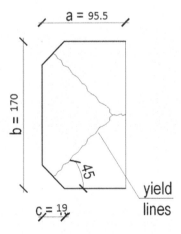

Figure 2.43 Internal bearing plate yield line pattern (fixed condition)

Therefore, having the dimensions a and b from Figure 2.43, the required bearing plate thickness t_p from the equation given in Section 2.6 is

$$t_p \geq \frac{2.8 \cdot a}{a + 0.9 \cdot b} \sqrt{\frac{2 \cdot a^2 \cdot w_u \cdot (3 \cdot b - 2 \cdot a)}{3 \cdot \phi \cdot f_{ay} \cdot (6 \cdot a + b)}} = 9.23 \, \text{mm}$$

So, the thickness t_p can be chosen: $t_p = 10$ mm.

The minimum recommended spacing between two plates is: $s_{p\_min} = 2a + t_p = 201$ mm. The maximum spacing allowed is $s_{p\_max} = 6a = 573$ mm. Calculating the required number of plates along the height of the column, $L/n_{plates} = 834$ mm, leads to best configuration using ten plates on each side of the profile with a spacing s_p equal to 500 mm. The weld throat required is $a_w = 5$ mm, which also meets the condition of $2a_w > t_p = 10$ mm, as shown in Figure 2.44.

The concrete compression struts are formed assuming an angle $\theta = 45°$, as shown in Figure 2.42. The resulting tie force in the transverse hoops then is: $F_{tie} = V_{Rd,1plate} = 181.9 \, kN$.

The necessary tie resistance can therefore be achieved for instance by combining stirrups $\Phi12$ with a spacing of 500 mm with $\Phi10$ with a spacing of 250 mm.

According to EN 1992-1-1 8.4.3(2), the basic required anchorage length $l_{b,rqd}$ is:

$$l_{b,rqd} = \frac{\phi}{4} \cdot \frac{\sigma_{sd}}{f_{bd}} = \frac{12 \, mm}{4} \cdot \frac{500 \, MPa}{4.5 \, MPa} = 333 \, \text{mm}$$

where: $f_{bd} = 2.25 \cdot \eta_1 \cdot \eta_2 \cdot f_{ctd} = 4.5 \, MPa$ is obtained with

$\eta_1 = 1$, coefficient related to the quality of the bond condition;
$\eta_2 = 1$, for $\Phi < 32$ mm;

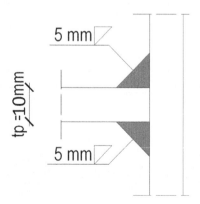

Figure 2.44 Weld dimensions of the plate connector

$$f_{ctd} = \frac{\alpha_{ctd} \cdot f_{ctk,0.05}}{\gamma_c} = \frac{1 \cdot 2.0 \, MPa}{1} = 2.0 \, MPa,$$ the value of design tensile strength;

$\sigma_{sd} = 500 \, MPa$, design stress of the bar;

The longitudinal reinforcement is considered as 4 $\Phi15$, like in the other options. The total amount of longitudinal reinforcement is $A_{sl} = 4 \cdot A_{\phi12} = 4.52 \, cm^2$, with a distance between bars equal to 345 mm.

Option 3 – Load transfer by a combination of shear studs and welded plate connectors

This solution combines flexible and stiff connectors (solutions 1 and 2) as shown in Figure 2.45a.

It is considered that the longitudinal shear is transferred by both the shear studs and the plate connectors adding the contributions of both types, as explained in Section 2.5. From the previous two options, it is already known that, for the plate connectors, $V_{Rd,1plate} = 181.9 \, kN$ and, for the shear studs, $P_{Rk} = 60$ kN. Using these values, a first estimate of the number of plate connectors and studs can be obtained as follows assuming that the load is equally distributed between both types of connectors.

$$\frac{N_{max}/2}{V_{Rd,1plate}} = 6.87 \text{ ; so, 8 plates (4 on each side)}$$

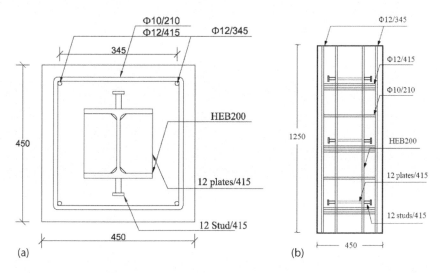

Figure 2.45 a, b Solution 3 – using both studs and plate connectors (a) top view (b) side view

$$\frac{N_{max}/2}{P_{Rk}} = 20.9 \; ; \text{ so, 22 studs (11 on each side)}$$

However, in order to improve the development of the strut-and-tie mechanism and the distribution of associated tying reinforcements, it is felt more appropriate to have the tying forces in the transverse reinforcement developing at the same level, i.e. finding a suitable solution involving the same number of studs and of plate connectors. In this way, a given hoop can be activated as a tie to anchor the compression struts generated by two plate and two studs.

For instance, with $n_{plates} = n_{studs} = 12$, the resistance of connectors is:

$$N_{Rd} = n_{plates} \cdot V_{Rd,1plate} + n_{studs} \cdot P_{Rk} = 2903 \text{kN}$$

The position of the connectors is shown in Figure 2.45(b). The distance between two rows of connectors is equal to $s_{pc} = 415$ mm.

The strut-and-tie model needed to determine the proper amount of transverse tying reinforcement is shown in Figure 2.46. The compression struts are formed with an angle, $\theta = 45°$.

The shear studs tie force is: $T_1 = \dfrac{P_{Rk}}{2} = 30 \text{kN}$ and $A_1 = \dfrac{T_1}{f_{sd}} = 0.60 \text{cm}^2$;

The plate connectors tie force is: $T_2 = V_{Rd,1plate} = 181.9 \text{kN}$ and $A_2 = \dfrac{T_2}{f_{sd}} = 3.63 \text{cm}^2$;

with a spacing $s_{pc} = 415$ mm.

Two solutions are therefore practically implementable to reach the required steel quantity: $2\Phi12$ ($A_s = 2.26$ cm²) @ 450 mm spacing or $2\Phi10$ ($A_s = 1.57 \text{cm}^2$) @ 225 mm spacing.

Vertical reinforcements are 4 $\Phi15$ like in the other solutions.

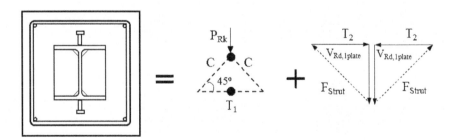

Figure 2.46 Connectors position along the cross-section

The total amount of vertical reinforcement is $A_{sl} = 4 \cdot A_{\phi 12} = 4.52\,cm^2$ at a distance of 345 mm between bars.

Option 4 – Load transfer by steel-concrete bond resistance
The solution evaluated in this case presents the simplest one among all four options from a design perspective, as no mechanical connectors are placed at the steel-concrete interface.

The following properties are assumed to design this solution (Figure 2.47(a)):

- The major axis of the steel profile is aligned perpendicularly to a column face.
- The vertical reinforcements (4Φ12 rebars) and the horizontal reinforcements (2Φ10 rebars) are considered similar to the previous configurations.
- The longitudinal shear is transferred homogenously on the total height of the column, with total embedded length of the profile l_{emb} equal to the total height of the column.

As an alternative to the values suggested in Table 2.2 that would not allow reaching the total required resistance, the bond stress is here evaluated according to Roeder et al. (1999), considering that the steel profile is not covered with rust or paint.

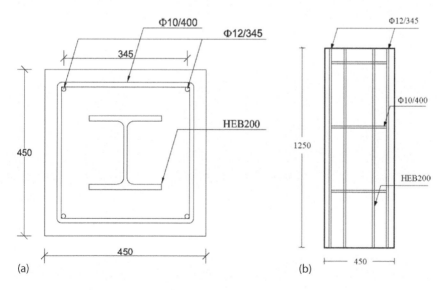

Figure 2.47 a, b Solution 4 – (a) top view and (b) side view

Two ratios are set as the primary variables affecting the bond stress capacity:

$$\frac{l_{emb\_HEB120}}{b_{HEB120}} = \frac{5000\,mm}{200\,mm} = 25 \text{ and} \rho = \frac{A_{HE120B}}{A_t} = 0.038$$

with $A_t = D * B = 450\text{ mm} * 450\text{ mm} = 202500\text{ mm}^2$ as total cross-section

No bond is available if $\frac{l_{emb\_HEB120}}{b_{HEB120}} \geq 0.125$.

The bond stress capacity can then be calculated as:

$$f_{B2\sigma} = 1.256 - 9.554 \cdot \rho = 0.893\,MPa$$

The perimeter of HE 200B is $P_{HEB200} = 1151$ mm.

The bond strength is equal to $f_{B2\sigma} \cdot P_{HE120B} \cdot l_{emb\_HE120B} = 5139.215\,kN$.

This is sufficient to transfer the total load of 2500 kN.

REFERENCES

ACI-318. (1995) Building code requirements for structural concrete, ACI 318-95. American Concrete Institute, Farmington Hills, MI.

AISC Standard 360-5. (2005) Specifications for structural steel buildings. American Institute for Steel Construction, Chicago, IL.

ACI-Committee-318. (1995) Commentary, ACI 318R-95. American Concrete Institute, Farmington Hills, MI.

AISC Standard 341–10. (2010) Seismic provisions for structural steel buildings. American Institute of Steel Construction, Chicago, IL.

Ali, M.S., Oehlers, D.J., and Bradford, M.A. (2001) Shear peeling of steel plates bonded to tension faces of RC beams. *Journal of Structural Engineering*, 127(12), 1453–1459.

Brown, C. (1966) Bond failure between steel and concrete. *The Journal of the Franklin Institute*, 182(5), 271–290.

Bryson, J., and Mathey R. (1962) Surface condition effect on bond strength of steel beams embedded in concrete. *ACI Materials & Structural Journals*, 59(3), 397–406.

CEN (European Committee for Standardization). (2004) Eurocode 2: Design of concrete structures, part 1.1 – General Rules for buildings. European Committee for Standardizations, Brussels.

Chen, J., and Teng, J. (2001) Anchorage strength models for FRP and steel plates bonded to concrete. *Journal of Structural Engineering*, 127(7), 784–791.

Daniels, B.J., and Crisinel, M. (1993) Composite slab behavior and strength analysis. Part II: Comparisons with test results and parametric analysis. *ASCE Journal of Structural Engineering*, 119(1), 36–49.

De Ville, V., Somja, H., and Pesesse, C. (2016) Finelg user's manual, version 9.1, Liège, bureau d'études greisch.

Degee, H., Plumier, A., Mihaylov, B., Dragan, D., Bogdan, T., Popa, N., Mengeot, P., Somja, H., De Bel, J.-M., Elghazouli, A., Bompa, D., Hjiaj, M., Nguyen, Q.-H. (2018) Smart composite components – concrete structures reinforced by steel profiles (SMARTCOCO). Final Report. European Commission. Research Fund for Coal and Steel. Grant Agreement RFSR-CT-2012-00031. ISSN 1831-9424 (PDF version).

Eurocode 2. (2004) Design of concrete structures, part 1.1 – General rules for buildings. European Committee for Standardizations, Brussels.

Eurocode 4. (2004) Design of composite steel and concrete structures, part 1.1– general rules for buildings. European Committee for Standardizations, Brussels.

Eurocode 8. (2004) Design provisions for earthquake resistance - part 1: General rules, seismic actions and rules for buildings. European Committee for Standardization, Brussels.

Fib. (2011a) Structural concrete: Textbook on behaviour, design and performance. Updated knowledge of the CEB/FIP model code 1990. fib Bulletin: 3.

Fib. (2011b) Design examples for strut-and-tie models. fib Bulletin: 61.

Hajjar, J.F., Schiller, P.H., and Molodan, A. (1998) A distributed plasticity model for concrete-filled steel tube beam-columns with interlayer slip. *Engineering Structures*, 20(8), 663–676.

Hamdan, M., and Hunaiti, Y. (1991) Factors affecting bond strength in composite columns. 3rd International Conference on Steel-Concrete Composite Structures, Fukuoka, Japan.

Hawkins, N. (1973) Strength of concrete encased steel beams. *Institution of Engineers (Australia), Civil Eng Trans*, CE15(1–2), 29–45.

Hawkins, N., and Mitchell, D. (1984) Seismic response of composite shear connections. *Journal of Structural Engineering*, 110(9), 2120–2136.

Johnson, R.P., and Oehlers, D.J. (1981) Analysis and design for longitudinal shear in composite T-beams. *Proceedings of the Institution of Civil Engineers*, 71(2), 989–1021.

Lungershausen, H. (1988) Zur Schubtragfähigkeit von Kopfbolzendübeln, Institut für Konstruktiven Ingenieurbau, Ruhr-Universität.

Lu, Y., and Kennedy, D. (1994) The flexural behaviour of concrete-filled hollow structural sections. *Canadian Journal of Civil Engineering*, 21(1), 111–130.

Marie, F., and Somja, H. (2018) Strut-and-tie model for the support of steel beams crossing concrete beams. In Romero et al. (Eds.), *Proceedings of the 12th international conference on advances in steel-concrete composite structures*, ASCCS 2018, 27–29 June 2018. Universitat Politecnica de Valencia, Valencia, 177–182. http://dx.doi.org/10.4995/ASCCS2018.2018.7018.

Oehlers, D.J. (2001) Development of design rules for retrofitting by adhesive bonding or bolting either FRP or steel plates to RC beams or slabs in bridges and buildings. *Composites Part A: Applied Science and Manufacturing*, 32(9), 1345–1355.

Oehlers, D.J., and Park, S. (1992) Shear connectors in composite beams with longitudinally cracked slabs. *Journal of Structural Engineering*, 118(8), 2004–2022.

Ollgaard, J.G., Slutter, R.G., and Fisher, J.W. (1971) Shear strength of stud connectors in lightweight and normal weight concrete. *AISC Engineering Journal*, 8(2), 55–64.

Plumier, A., Bogdan, T., and Degee, H. (2013) Design for shear of columns with several encased steel profiles. Internal Report. University of Liege – ArcelorMittal [Available on request].

Plumier, A. (2016) SMARTCOCO project deliverable 2.1: State of the art on force transfer mechanisms at steel-concrete interface behavior of hybrid structural components. *Plumiecs Report* [Available on request].

Rabbat, B., and Russell, H. (1985) Friction coefficient of steel on concrete or grout. *Journal of Structural Engineering*, 111(3), 505–515.

Roeder, C. (1984) Bond stress of embedded steel shapes in concrete. *Proceedings of the U.S./Japan Joint Seminar.*

Roeder, C., Chmielowski, R., and Brown, C.B. (1999) Shear connector requirements for embedded steel sections. *Journal of Structural Engineering*, 125(2), 142–151.

Salari, M.R. (1999) *Modeling of bond-slip in steel-concrete composite beams and reinforcing bars.* Doctoral Thesis. Department of Civil, Environmental and Architectural Engineering, University of Colorado at Boulder ProQuest Dissertations Publishing, UMI Number: 9955312. Bell and Howell Information and learning Company, 300 North Zeeb Road, Ann Arbor, MI 48106-1346

Schlaich, J., Schafer, K., and Jennewein, M. (1987) Toward a consistent design of structural concrete. *PCI Journal*, 32(3), 74–150.

Shakir-Khalil, H. (1993a) Pushout strength of concrete-filled steel hollow sections. *The Structural Engineer*, 71(13), 230–233.

Shakir-Khalil, H. (1993b) Resistance of concrete-filled steel tubes to pushout forces. *The Structural Engineer*, 71(13), 234–243.

Viest, I., Griffs, L.G., and Wyllie, L.A. Jr. (1997) *Composite construction design for buildings.* ASCE: McGraw-Hill, New York.

Wium, J., and Lebet, J. (1991) *Composite columns: Force transfer from steel section to concrete encasement.* Ecole Polytechnique Federale, Lausanne.

Wium, J., and Lebet, J. (1992) *Force transfer in composite columns consisting of embedded HEB 300 and HEB 400 sections.* Ecole Polytechnique Federale, Lausanne.

Wium, J., and Lebet, J. (1994) Simplified calculation method for force transfer in composite columns. *Journal of Structural Engineering*, 120(3), 728–746.

Womersley, W. (1927) Bond between concrete and steel. *Concrete and Construction Engineering*, 22(2), 153–159.

Chapter 3

Analysis of hybrid steel-concrete structures and components

Hugues Somja and Pisey Keo

CONTENTS

DOI: 10.1201/9781003149811-3

3.1 INTRODUCTION

Structural analysis is usually performed to determine deformations and internal forces in structural components. Those deformations and internal forces can be used later to verify the global and local deflections and the individual resistance of the components by using the design guides or design codes for the type of structure being analyzed. For composite steel-concrete structures, the global structural analysis should be in accordance with composite steel-concrete structural design codes. Similarly, the global structural analysis should be generally performed in accordance with reinforced concrete structural design codes for structures where the structural behaviour is essentially the one of reinforced or pre-stressed concrete structures. In contrast to well-established design codes for reinforced concrete and composite steel-concrete structures, the design of hybrid structures is less documented. Therefore, a design guide for hybrid structures has been proposed through the European SMARTCOCO project. This chapter aims to present the results of the SMARTCOCO project related to the analysis of hybrid structures and components. In terms of analysis models, four methods have been proposed: linear elastic analysis, first-order elastic analysis with amplification factors, second-order elastic analysis and nonlinear analysis. The choice of the method is based on the complexity of the hybrid structures or components and on their sensitivity to second-order effects.

The rest of the chapter is organized as follows. Primarily, the scope of the design guides is highlighted in Section 3.2. Section 3.3 presents recommendations for the analysis of structures and components based on linear and

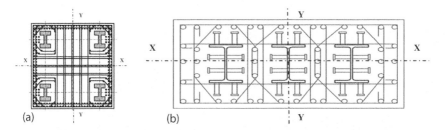

Figure 3.1 (a) Hybrid cross-section with frour embedded steel sections. (b) Hybrid cross-section with three embedded steel sections

nonlinear analysis, including the simplified method for analyzing slender hybrid components. In Section 3.4, the background of the simplified method is described. Section 3.5 is dedicated to the software developed as part of the SMARTCOCO project for analyzing the hybrid components subjected to axial load and uniaxial bending moment. The chapter closes by presenting a design example of a slender hybrid column using the recommended simplified method.

3.2 SCOPE

The scope of the guidelines proposed here is limited to structures containing hybrid components and to hybrid components, whose cross-section is doubly symmetric and uniform over the component length. A hybrid cross-section may have more than one embedded steel section for the bending direction considered, as for example the column around bending axis XX and YY depicted in Figure 3.1a or the wall bending around axis YY depicted in Figure 3.1b.

It is worth noting that the wall bending around axis XX illustrated in Figure 3.1b should be checked with the rules of composite structures. In addition, the lateral torsional buckling resistance of the wall bending around axis YY as shown in Figure 3.1b should be verified according to Section 3.3.7. Shear effects are not considered in the design of hybrid components in this chapter. The reader is invited to refer to Chapter 4 for shear action effects in hybrid components.

3.3 ANALYSIS OF STRUCTURES AND HYBRID COMPONENTS

3.3.1 General introduction

Global structural analysis of structures is performed to determine deformations and internal forces in structural components. Adequate results are obtained when the analysis type, the geometry (including imperfections) and the stiffness of each component are defined appropriately. The structural analysis should conform with the design codes for the type of structures being analyzed. For composite steel-concrete structures consisting of hybrid components and composite or structural steel components and joints, the global structural analysis should be in accordance with composite steel-concrete structural design codes, for example, Eurocode 4 (2005a). Similarly, the global structural analysis should be generally based on reinforced concrete structural design codes for structures where the structural behaviour is essentially one of reinforced or pre-stressed concrete structures with only a few composite or hybrid components. For both types of structures, an appropriate stiffness of the hybrid component should be used in the global structural analysis.

In terms of an analysis model, four methods can be used: linear elastic, first-order elastic analysis with amplification factors, second-order elastic analysis and nonlinear analysis. The latter can be used in general cases. On the one hand, when the second-order effects on the global behaviour of the structure can be ignored, the linear elastic model can be implemented to determine the internal forces in the structural components. The structural components can be then verified individually for their cross-section resistance and their resistance to instability using those internal forces. On the other hand, when the second-order effects on the global behaviour of the structure cannot be ignored, second-order elastic analysis or first-order elastic analysis with amplification factors should be used.

It is noteworthy that in the design of slender structures, the second-order effects need to be considered. European codes provide guidance on how to consider these effects in structural analysis using either a precise second-order analysis or a simpler but less precise first-order analysis with appropriate amplification factors. Nevertheless, second-order effects may be ignored if they are significantly lesser than the corresponding first-order ones (normally if less than 10% of first-order effects). This implies that the designer would first check the second-order effects before ignoring them. However, European codes provide simplified criteria to verify if the second-order effects must be taken into account in global structural analysis. For instance, Eurocode 4 (2005a) proposes to evaluate α_{cr}, the factor by which the design loading would have to be increased to cause elastic instability. If the aforementioned factor is greater than 10, the second-order effects can be ignored. Eurocode 2 (2005a) proposes to evaluate the geometric slenderness of the component and total vertical loads in the structure as a whole to assess the second-order effects for the isolated component and the structure, respectively. The second-order effects can be ignored if the aforementioned parameter (for isolated components or structures) is less than a certain limited value. If the second-order effects must be considered, Eurocode 2 (2005a) refers to Appendix H for the evaluation of the global second-order effects using a first-order elastic analysis with magnified horizontal forces, where the rigidity of bracing elements is determined by taking into account concrete cracking. Components sensitive to second-order effects will then be checked separately using the internal forces given by the global structural analysis. Likewise, Eurocode 4 (2005a) recommends using a first-order analysis with appropriate amplification in order to include indirect second-order effects. Moreover, Eurocode 4 (2005a) states that individual stability checks of composite columns can be ignored if their individual imperfection and their nominal stiffness are fully accounted for in the global structural analysis including global second-order effects.

First-order analysis with amplification factors is generally known as the moment magnification method when being applied to structural components. It can be written in general form as $M_{Ed,2} = k \times M_{Ed,1}$, where $M_{Ed,2}$ is the evaluation of the second-order bending moment; $M_{Ed,1}$ is the first-order bending moment; and k is the so-called moment magnification factor. A large number of equations for k proposed in the technical literature can be (re)written in the following form: $k = \beta/(1 - N_{Ed}/N_{cr})$ where N_{Ed} is the design axial load; N_{cr} is the elastic critical normal force; and β is the equivalent uniform moment factor. The accuracy of the moment magnification method strongly depends on, as included in N_{cr}, the effective flexural stiffness EI of the component which depends on, among other factors, the nonlinearity of the concrete stress-strain curve, the creep and the cracking along the component length, and on the factor β. It can be seen that the effective flexural stiffness EI for components or structures sensitive to second-order effects needs to be known in advance in order that first-order elastic analysis with amplification factors can be performed.

The same approach can be adopted for structures possessing hybrid components whose cross-section types are defined in Section 3.2 (Figure 3.1). This means that:

- If second-order elastic analysis is performed, and the individual imperfection and the nominal stiffness of hybrid components are introduced, the verification of resistance to buckling of those hybrid components is not required. However, the resistance of cross-sections has to be verified by using the action effects obtained from the global second-order elastic analysis.
- If first-order elastic analysis with amplification factors is performed and the nominal stiffness of hybrid components is introduced, the verification of resistance to buckling of those hybrid components is required and has to be verified by using the action effects obtained from global structural analysis.
- On the other hand, by means of an appropriate simplified method, for instance the moment magnification method, hybrid components sensitive to second-order effects can be checked separately using internal forces obtained from global structural analysis.

3.3.2 Nonlinear analysis

In general, nonlinear analysis can be performed to study the behaviour of structures possessing hybrid components. In that case, the geometric and material nonlinearities (long-term effects included) as well as local and global imperfections should be considered. Moreover, the effects of slip and separation (uplift) on the calculation of internal forces and moments at interfaces between steel and concrete should be taken into account.

3.3.2.1 Global geometric imperfections

For composite steel-concrete structures consisting of hybrid components and composite or structural steel components and joints, the global imperfections of the structure should be in accordance with composite steel-concrete structural design codes, such as Eurocode 4 (2005a). Similarly, the global imperfections of the structure should be generally based on reinforced concrete structural design codes for structures where the structural behaviour is essentially one of reinforced or pre-stressed concrete with only a few composite or hybrid components.

3.3.2.2 Hybrid component imperfections

The action effects at component ends are generally influenced by the global sway imperfections and insignificantly by the local bow imperfections. Hence, the effects of component imperfections on end moments and forces can be ignored in global structural analysis if the axial load ratio is less significant, for instance in the case that the design axial load N_{Ed} is lower than 25% of the Euler critical load defined as $N_{cr} = \pi^2 (EI)_{eff,II} / L^2$, in which L is the effective length of the component and $(EI)_{eff,II}$ is the effective flexural stiffness of the component, defined later in Section 3.3.3. However, the component imperfections should always be considered in the global second-order analysis when verifying the stability of the compression component within its length. It is well known that the definition of the initial imperfections strongly affects the behaviour of slender components. For concrete columns, Eurocode 2 (2005a) recommends considering a geometric imperfection equal to $L/400$, whereas for steel columns, Eurocode 3 (2005a) suggests adopting a geometric imperfection equal to $L/1000$ but also taking into account the effects of the residual stresses on steel cross-sections. For composite columns, the imperfect shape is governed by steel components and therefore by the residual stress distribution. Accordingly, an initial imperfection $L/1000$ has to be considered and the geometric effects of the residual stress distribution must be considered. To simplify the calculation, Eurocode 4 (2005a) proposes replacing the residual stresses with an equivalent initial bow imperfection. However, Bergmann and Hanswille (2006) have shown that this simplification produces an approximate value of the ultimate resistance in axial compression. For hybrid components, since steel components are embedded in concrete, we can adopt an initial imperfection of $L/400$ to concrete components and $L/1000$ with residual stresses to steel components. However, considering different initial imperfections in different components for a structural component is more likely to be complicated. Hence, it is here proposed to consider an initial bow imperfection w_0 equal to $L/400$ without considering residual stresses for hybrid compression components. Indeed, according to the parametric study (Keo et al., 2015b), which will be presented in the background of the simplified method, the effects of residual stresses are negligible for slender hybrid columns. Thus, this consideration would give a reasonable conservative result.

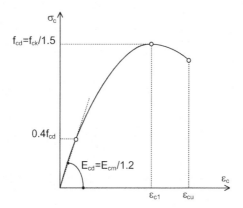

Figure 3.2 Design of concrete stress-strain curve

3.3.2.3 Material constitutive model

Even if Eurocodes are organized in separate design codes according to materials used in the structure, a consistent set of material models is proposed for nonlinear modelling. The stress-strain models with partial coefficients depicted in Figure 3.2 to Figure 3.4 can thus be adopted for hybrid structures. It is worth mentioning that a strain-hardening ratio of steel reinforcement of 1/300 is recommended.

The stress-strain relationships for concrete shown in Figure 3.2 are given by Equation (3.1).

$$\frac{\sigma}{f_{cd}} = \frac{k\eta - \eta^2}{1 + (k-2)\eta} \qquad (3.1)$$

where:

- $\eta = \varepsilon_c / \varepsilon_{c1}$
- $k = 1.05 \, E_{cd} \times |\varepsilon_{c1}| / f_{cd}$
- Equation (3.1) is valid only for $0 < |\varepsilon_c| < |\varepsilon_{cu}|$.

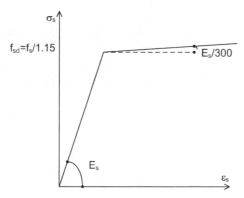

Figure 3.3 Design of reinforcement stress-strain curve

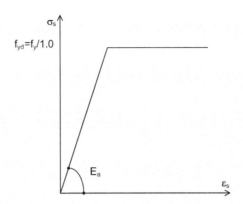

Figure 3.4 Design of structural steel stress-strain curve

The concrete creep can be taken into account by dividing both the design value of modulus of elasticity and concrete strain by a factor $(1+\varphi_{ef})$, in which φ_{ef} is the effective creep ratio.

The following symbols are adopted in Figure 3.2 to Figure 3.4:

- f_{ck}: characteristic compressive cylinder strength of concrete at 28 days.
- f_{cd}: design value of concrete compressive strength.
- E_{cm}: secant modulus of elasticity of concrete.
- E_{cd}: design value of modulus of elasticity of concrete.
- ε_{c1}: compressive strain of concrete at the peak stress.
- ε_{cu}: ultimate compressive strain of concrete.
- f_s: characteristic value of the yield strength of reinforcing steel.
- f_{sd}: design value of the yield strength of reinforcing steel.
- E_s: design value of modulus of elasticity of reinforcing steel.
- f_y: characteristic value of the yield strength of structural steel.
- f_{yd}: design value of the yield strength of structural steel.
- E_a: design value of modulus of elasticity of structural steel.

3.3.3 Linear elastic analysis

To determine the action effects, linear elastic analysis should be carried out assuming cracked concrete cross-sections, linear stress-strain relationships and elastic modulus taking into account creep strains. It is worth noting that concrete standards (Eurocode 2, 2005a) often offer the possibility to consider, for the distribution of internal forces, all reinforced concrete cross-sections un-cracked in the linear elastic analysis by using a mean value of the instantaneous modulus of elasticity. However, assuming an un-cracked concrete cross-section as well as without considering delayed strains may overestimate the flexural stiffness of the components with the least reinforcement. This results in an artificial distortion of the structure and an unintended redistribution of forces to the less resistant components.

To determine the internal force in a first-order analysis, the design value of effective flexural stiffness $(EI)_{eff,II}$ of hybrid components taking into account concrete cracking may be taken as:

$$(EI)_{eff,II} = K_0 \left(E_a I_a + E_s I_s + K_{e,II} E_{cm} I_c \right) \tag{3.2}$$

where:

- E_a, E_s and E_{cm} are the mean values of the modulus of elasticity of structural steel, reinforcement and concrete, respectively.
- I_a, I_s and I_c are the second moments of area of structural steel, reinforcement and concrete cross-section.

The constants K_0 and $K_{e,II}$ are calibration and correction factors which should be taken as 0.9 and 0.5, respectively. These factors are the same as those defined in Eurocode 4 (2005a). It is worth mentioning that according to the parametric study (Keo et al., 2015b), using the effective flexural stiffness defined by Equation (3.2) in the simplified method, based on the moment magnification method for isolate hybrid column design, gives a satisfactory result. Despite a maximum derivation of around 20% for hybrid columns having low or moderate slenderness, a good mean value of column bearing capacity compared to the one given by nonlinear finite element analysis is obtained. Accordingly, Equation (3.2) can be used in global first-order analysis to estimate the internal forces.

In addition, the influence of long-term effects on the effective elastic flexural stiffness should be taken into account in Equation (3.2) by reducing the elastic modulus of concrete E_{cm} to:

$$E_c = E_{cm} / (1 + N_{G,Ed}/N_{Ed}\, \varphi_t) \text{ or } E_c = E_{cm} / (1 + M_{G,Ed}/M_{Ed}\, \varphi_t) \tag{3.3}$$

where φ_t is the creep coefficient as defined in Eurocode 4 (2005a) (equivalent to $\varphi(t,t_0)$ defined in Eurocode 2 (2005a)); N_{Ed} or M_{Ed} is the total design normal force or bending moment, respectively; and $N_{G,Ed}$ or $M_{G,Ed}$ is a part of the total design normal force or bending moment, respectively, that is permanent.

3.3.4 First-order elastic analysis with amplification factors

The first-order elastic analysis with amplification factors is an approach based on magnification of the first-order actions. This method is allowed to be used by Eurocode 4 (2005a); however, no further information is given.

Considering for example the structure subjected to gravity loads $F_{V,Ed}$ and to the first-order horizontal forces $F_{H,0Ed}$ due to wind and which include the equivalent loads ($F_{V,Ed,0}$) due to initial out-of-plumbness θ, the second-order effects in the structure can be taken into account by magnifying the horizontal forces and with a factor k, so-called magnification factor. The latter can be expressed in different ways. For reinforced concrete buildings, Appendix H of Eurocode 2 (2005a) gives the following equation for determining the magnification factor:

$$k = \frac{1}{1 - \dfrac{F_{V,Ed}}{F_{V,B}}} \tag{3.4}$$

in which $F_{V,B}$ is the nominal global buckling load which is calculated with nominal stiffness of the bracing components and taking concrete cracking into account. The detailed formulation of $F_{V,B}$ can be found in Appendix H of Eurocode 2 (2005a).

The first-order elastic analysis with amplification factors described above can also be applied to structures possessing hybrid components. In this case, the nominal stiffness of slender hybrid compression components should be defined in such a way that the action effects obtained with the global first-order analysis with amplification factors can be used directly to check the resistance of the cross-section. A parametric study using the nonlinear finite element model has been performed, and the following equation used to estimate the nominal stiffness $(EI)_{eff,II}$ of slender hybrid compression components with cross-sections defined in Section 3.2 has been proposed (Keo et al., 2015b):

$$(EI)_{eff,II} = K_c E_{cd} I_c + K_{sa} E_s I_s + K_{sa} E_a I_a \tag{3.5}$$

where:

- E_{cd} is the design value of the modulus of elasticity of concrete defined as $E_{cd} = E_{cm} / \gamma_{cE}$ in which γ_{cE} is recommended to be taken equal to 1.2.
- K_c is a factor for effects of cracking, creep, etc., given by $K_c = k_1 k_2 / (1 + \varphi_{ef})$.
- K_{sa} is a factor for contribution of reinforcement, defined as:

$$K_{sa} = \begin{cases} 1, & \text{if } \rho_{sa} = \dfrac{A_s + A_a}{A_c} \le 0.04 \\[2ex] \dfrac{0.450 \lambda^{0.147}}{1 + 1.433 \varphi_{ef} exp(-0.027\lambda)}, & \text{if } \rho_{sa} = \dfrac{A_s + A_a}{A_c} > 0.04 \end{cases}$$

- φ_{ef} is the effective creep ratio.

- $k_1 = (f_{ck} / 20)^{0.5}$ is a factor which depends on concrete strength class in MPa.
- $k_2 = n\,\lambda\,/\,170 \leq 0.2$ is a factor which depends on axial force ratio, $n = N_{Ed}/N_{pl,Rd}$, and slenderness ratio $\lambda = L\,/\,(I_g/A_g)^{0.5}$ where I_g and A_g are the moment of inertia and the area of the hybrid compression component gross cross-section considered un-cracked and L is the effective length of the compression component.

It is worth mentioning that the development of the equation of effective flexural stiffness $(EI)_{eff,II}$ is based on the one proposed by Eurocode 2 (2005a). The development details are presented in the section devoted to the background of the simplified method. In fact, it is shown that for a low percentage of steel reinforcement in cross-section, the hybrid column can be considered as a reinforced concrete one, since residual stresses in steel sections have nearly no influence on the column behaviour, and their individual moment of inertia is insignificant compared to the one of hybrid section. Hence, the correction factors proposed in Eurocode 2 (2005a) can be adopted. However, due to the dependence of steel yielding to the column slenderness and concrete creep, a new correction factor of steel components is proposed for a high percentage of steel reinforcement in cross-section.

3.3.5 Second-order elastic analysis based on nominal stiffness

In a second-order elastic analysis based on stiffness, nominal values of the flexural stiffness to be used for components sensitive to second-order effects should take into account the effects of cracking, material nonlinearity and creep on the overall behaviour of those components. This also applies to adjacent components involved in the analysis, e.g. beams, slabs or foundations. Where relevant, soil-structure interaction should be taken into account. The nominal stiffness should be defined in such a way that the total bending moments resulting from the analysis can be used to check the resistance of cross-sections to bending moment and axial force.

3.3.5.1 Geometric imperfections

The global and local imperfections described in Section 3.3.2(a) and 3.3.2(b) should be adopted.

3.3.5.2 Nominal stiffness

For hybrid compression components with cross-sections defined in Section 3.2, Equation (3.5) should be used to determine the nominal stiffness $(EI)_{eff,II}$.

3.3.6 Simplified method for slender hybrid components

In the moment magnification method, the criterion to verify the resistance of the compression element is that the design (second-order) bending moment M_{Ed2} has to be lesser than or equal to the plastic bending moment of the cross-section $M_{pl,N,Rd}$ taking into account the applied normal force.

3.3.6.1 Resistance of cross-section

To determine $M_{pl,N,Rd}$ as well as the M-N interaction curve of a fully plastic cross-section, the compatibility strain method may be used, as it has been validated for the hybrid section by Bogdan et al. (2012). For this method, the following assumptions are made:

- Plane section remains plane.
- Interlayer slip between steel and concrete is ignored.
- Parabola rectangle stress-strain relationship of concrete is adopted as proposed by Eurocode 2 (2005a).
- Bilinear law of steel is used.

The possible strain distribution in the ultimate limit state of the hybrid column with three pivots named A, B and C is shown in Figure 3.5. Pivot A represents the distribution where the reinforcement bars at the bottom reach the strain limit in tension (ε_{ud} =10‰) while the top fibre varies from tension to the ultimate limit strain of concrete in compression (ε_{cu2} =3.50‰) which represents point B. This point is called pivot B. The last pivot, C, is regarded as the strain limit in pure compression of concrete (ε_{c2} =2.00‰).

3.3.6.2 Magnification factor

The design moment M_{Ed2} in a slender compression element is evaluated by a magnification of the first-order bending moment M_{Ed1}, where M_{Ed1} results from the calculated bending moment M_{Ed} and axial force N_{Ed} obtained by a linear elastic analysis and from the geometrical imperfection (w_0) of the slender element, i.e. $M_{Ed1} = M_{Ed} + N_{Ed} \times w_0$.

The magnification factor k is defined by:

$$k = \frac{\beta}{1 - \dfrac{N_{Ed}}{N_{cr,eff}}} \tag{3.6}$$

where β=0.6+0.4 $r_m \geq$ 0.4 in which r_m is the ratio of end moments. The critical buckling load $N_{cr,eff}$ of the slender element is equal to $\pi^2(EI)_{eff,II}/L^2$ where L is the effective length of the compression element. The equation of effective flexural stiffness $(EI)_{eff,II}$ defined in Equation (3.5) may be used.

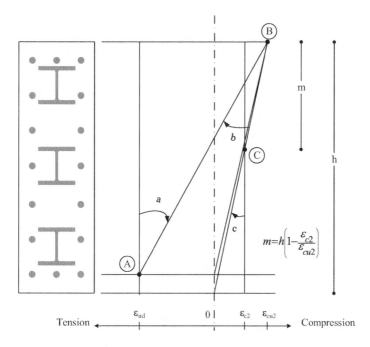

Figure 3.5 Possible strain distribution in ultimate limit state of a hybrid column

3.3.7 Lateral torsional buckling

For components consisting of slender hybrid cross-sections (hybrid components with three embedded steel profiles for example), the resistance to lateral torsional buckling should be verified by considering that the compressive part of the hybrid cross-section, see Figure 3.6, will buckle in the out-of-plane direction. The axial load to be used can be determined by integrating the compressive stress generated by compressive force and by magnified bending moment on the compressive section.

3.4 BACKGROUND OF THE SECOND-ORDER ELASTIC ANALYSIS BASED ON NOMINAL STIFFNESS

3.4.1 Introduction

The second-order effects on the behaviour of hybrid components can be analyzed using a simplified method or a more advanced calculation tool such as nonlinear finite element analysis. Eurocode 2 (2005a) and Eurocode 4 (2005a) propose both simplified methods and general methods based on nonlinear second-order analysis where geometric and material nonlinearities are considered. Both codes propose simplified design methods based on

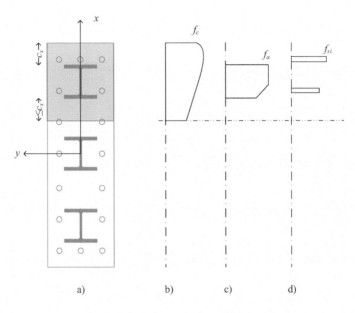

Figure 3.6 Cross-section and axial force considered for lateral torsional buckling. a) Cross-section considered to be buckled around x-axis. b) Possible stress distribution on concrete section. c) Possible stress distribution on steel section. d) Possible stress distribution in rebars

the moment magnification approach. The latter can be written in general form as $M_{Ed,2} = k \times M_{Ed,1}$, where $M_{Ed,2}$ is the evaluation of the second-order bending moment; $M_{Ed,1}$ is the first-order bending moment; and k is the so-called moment magnification factor. A large number of equations for k proposed in the technical literature can be (re)written in the following form: $k = \beta/(1 - N_{Ed}/N_{cr,eff})$ where β is the equivalent uniform moment factor. The accuracy of the moment magnification method strongly depends on, as included in $N_{cr,eff}$, the effective flexural stiffness $(EI)_{eff,II}$ which depends on, among other factors, the nonlinearity of the concrete stress-strain curve, the creep and the cracking along the column length. Since hybrid components are neither reinforced concrete structures in the sense of Eurocode 2 (2005a) nor composite steel-concrete structures in the sense of Eurocode 4 (2005a), the equation used for defining the effective flexural stiffness (EI) of reinforced concrete or composite columns cannot be straightforwardly used for one of the hybrid columns. In fact, according to a parametric study performed by Keo et al. (2015b), a straightforward application of the moment magnification method of Eurocode 2 and Eurocode 4 (with the corresponding effective flexural stiffness of reinforced concrete and composite columns) for hybrid column design leads to a wide scatter, where half of the case studies are unsafe when compared to nonlinear finite element results. Therefore, a new equation of effective flexural stiffness for hybrid columns to be used in the simplified method for second-order analysis of hybrid compression components is proposed by Keo et al. (2015b). The development of this new equation of effective flexural stiffness is based on an extensive parametric study achieved with nonlinear finite element analysis. The development of finite element formulation is based on the extension of a vast amount of research on the behaviour of composite beams with interlayer slips.

The following section aims to present the nonlinear finite element formulation for the large displacement analysis of hybrid planar beam/column with several encased steel profiles taking into account the slips occurring at each steel-concrete interface. The finite element model provides an efficient tool for nonlinear analysis of hybrid beam columns and serves as a reference for an extensive parametric study to develop a new equation of effective flexural stiffness for hybrid columns to be used in the simplified method for second-order analysis of hybrid compression components.

3.4.2 Finite element model

3.4.2.1 Finite element formulation

In order to analyze the behaviour of slender hybrid columns, a two-dimensional beam-column finite element formulation is developed based on Euler-Bernoulli kinematics (shear deformation is negligible) and fibre cross-section discretization. The latter permits to consideration of some 3D effects on uniaxial behaviour, like the confinement of concrete by using an appropriate stress-strain relationship. Furthermore, to consider geometric nonlinearity, the co-rotational approach is adopted. In this context, the element displacements are separated into rigid-body and deformational degrees of freedom. The element rigid-body motion is handled separately via the mapping from the co-rotational frame to the global coordinate system. In addition, the slips that may occur at the interfaces of embedded steels and concrete components are considered. The developed FE model is then capable to consider the following aspects: a cross-section with more than one steel section in partial interaction, geometrical and material nonlinearities, initial imperfection, residual stresses, and concrete confinement.

A brief description of the FE formulation is presented herein for the case of a hybrid column with two encased steel profiles. However, the concepts are also applicable to the general case of several encased steel profiles. A more detailed deduction can be found in Keo et al. (2015a).

Let us consider a planar element with two steel sections fully encased in concrete and including shear connectors at the contact interface uniformly distributed along the element length, as shown in Figure 3.7. It is assumed that the interlayer slip can occur at the interface but there is no uplift since the steel sections are embedded in concrete.

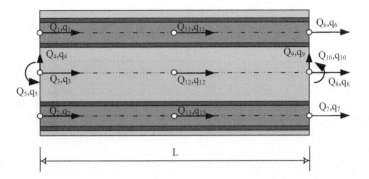

Figure 3.7 Degree of freedom of local linear element for two encased steel profiles

For the present case, the element is divided into three sub-elements: two for structural steel profiles (s_k, k=1, 2) and one for reinforced concrete (c). This division leads to ten global degrees of freedom in the fixed global coordinate system: global displacements and rotation of the nodes (c_i and c_j) and slips (g_{ki}, g_{kj}) between the steel node (s_{ki}, s_{kj}) and concrete node (c_i and c_j). Since all components are bent according to Euler-Bernoulli kinematics, the rotation of all components (steel sections and concrete section) at the end nodes are equal, and the slips (g_{ki}, g_{kj}) are perpendicular to the end cross-sections. The vector of global nodal displacements is defined by:

$$\mathbf{p}_g = \left[u_{ci}, v_{ci}, \theta_i, g_{1i}, g_{2i}, u_{cj}, v_{cj}, \theta_j, g_{1j}, g_{2j} \right]^{\mathrm{T}}$$

Due to the presence of the three rigid-body modes in the global coordinate system, the corresponding element stiffness matrix is singular. Therefore, the linear local element is derived in the local system (x_l, y_l) without rigid-body modes. The latter translates and rotates with the element as the deformation proceeds. In this local system, the element has seven degrees of freedom, and the vector of local displacements is defined as:

$$\mathbf{p}_l = \left[\bar{u}_{s1i}, \bar{u}_{s2i}, \bar{\theta}_i, \bar{u}_{s1j}, \bar{u}_{s2j}, \bar{u}_{cj}, \bar{\theta}_j \right]^{\mathrm{T}}$$

Without rigid-body modes, the local element has ten degrees of freedom (see Figure 3.8). The transverse displacement $\bar{v}$ is approximated using cubic Hermite interpolations. In addition, in order to avoid curvature locking, three internal nodes (one for each component) are added in order to use a quadratic shape function for axial displacement interpolation. However, for saving the calculation time, three degrees of freedom corresponding to the internal nodes will be statically condensed out thereafter to obtain the local displacement vector containing only the degrees of freedom at the element ends.

3.4.2.2 Hypothesis for finite element analysis for hybrid columns

Incremental finite element computations usually use the stress-strain relationships of the materials based on the design values of the strengths. The stiffness of the elements is then derived from these stress-strain curves. Since there is a close coherence between strength and stiffness, the different safety factors for steel, concrete and reinforcement must be carefully applied in the design using finite element models. In the general finite element method of Eurocode 2 (2005a), the stress-strain relationships based on the design values are clearly defined by the code. In contrast, regarding the safety concept in the case of nonlinear finite element analysis, the Eurocode for composite structures gives no clear guidance. Therefore, the safety format proposed by Eurocode 2 (2005a) is adopted. Furthermore, concerning column imperfections, since the hybrid column is fabricated as a reinforced concrete column, the initial imperfection w_0 equal to $L/400$ is adopted, and the residual stresses in structural steel are considered in the model.

The design stress-strain curves for each material are illustrated in Figures 3.2 to 3.4, and the distribution of residual stresses in structural steel profiles depicted in Figure 3.9 are adopted. The strain-hardening ratio of steel reinforcement can be taken as equal to 1/300, which is a choice calibrated to the experimental uniaxial test.

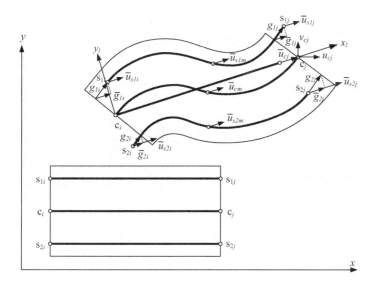

Figure 3.8 Co-rotational kinematics of hybrid beam with two encased steel profiles

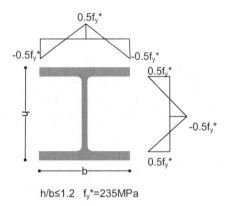

h/b≤1.2   f$_y$*=235MPa

Figure 3.9 Residual stress distribution

3.4.2.3 Validation of FE model

To the best of our knowledge, there are limited available experimental test results for buckling tests on RC columns with multi-embedded steel profiles (hybrid columns) in the technical literature. A couple of experimental compression-bending tests on steel-concrete shear walls with vertical steel-encased profiles were conducted by Dan et al. (2011) and by Zhou et al. (2010). Bogdan et al. (2019) conducted a series of experimental tests on six ¼ scaled concrete-encased composite columns with multiple separated steel sections subjected to an eccentric axial load. Three ratios of eccentricity (*e/D=0%, 10%* and *15%)* are considered in Bogdan et al. (2019). However, the dimension of the tested specimens in Dan et al. (2011), Zhou et al. (2010) and Bogdan et al. (2019) are such that they cannot be considered slender columns. Therefore, besides validating the developed finite element model by comparing its prediction against the six test results of (Bogdan et al., 2019), ten test results of eccentrically loaded slender composite columns (Al-Shahari et al., 2003; Morino et al., 1985) and six test results of short composite columns (Chen and Lin, 2006) are also considered. For the sake of clarity, in this study we denote six specimens tested by Bogdan et al. (2019) as MCSRS1–MCSRS6, seven specimens tested by Al-Shahari et al. (2003) as CESC1–CESC7, three specimens tested by Morino et al. (1985) as CESC8–CESC10 and six concrete-encased steel composite short columns tested by Chen and Lin (2006) as SCESC1–SCESC6. The geometrical and material properties of the above-mentioned specimens are summarized in Table 3.1.

All composite column specimens are pinned at both ends. The columns MCSRS1–MCSRS6 and CESC1–CESC10 are loaded with the same eccentricity at both extremities. The concrete region is subdivided into three parts as suggested by Mirza and Skrabek (1991). A highly confined concrete zone is taken from the web of the steel section to each flange, and a partially

Table 3.1 Specimen dimensions and material properties

Specimen	B × D (mm)	kL (mm)	Structural steel	Long. bar	e/D	f_c (MPa)	f_y (MPa)	f_s (MPa)
MCSRS1	450×450	2700	H120×106×12×20	3Ø8	0.0	61.17[a]	465[c]	438
MCSRS2	450×450	2700	H120×106×12×20	3Ø8	0.0	56.62[a]	404[c]	438
MCSRS3	450×450	2700	H120×106×12×20	3Ø8	0.1	59.75[a]	429[c]	438
MCSRS4	450×450	2700	H120×106×12×20	3Ø8	0.1	68.40[a]	399[c]	438
MCSRS5	450×450	2700	H120×106×12×20	3Ø8	0.15	67.50[a]	390[c]	438
MCSRS6	450×450	2700	H120×106×12×20	3Ø8	0.15	75.17[a]	397[c]	438
CESC1	230×230	2000	H100×96×5×8	4Ø12	0.3	20.5[a]	337	459
CESC2	230×230	2000	H100×96×5×8	4Ø12	0.3	13.7[a]	337	459
CESC3	230×230	2000	H140×133×5.5×8	4Ø12	0.3	20.5[a]	307	459
CESC4	230×230	2000	H140×133×5.5×8	4Ø12	0.3	28.2[a]	307	459
CESC5	230×230	3000	H140×133×5.5×8	4Ø12	0.3	28.2[a]	307	459
CESC6	230×230	3000	H100×96×5×8	4Ø12	0.17	20.5[a]	337	459
CESC7	230×230	3000	H100×96×5×8	4Ø12	0.17	13.7[a]	337	459
CESC8	160×160	960	H100×100×6×8	4Ø6	0.25	21.1[a]	345	460
CESC9	160×160	2400	H100×100×6×8	4Ø6	0.25	23.4[a]	345	460
CESC10	160×160	3600	H100×100×6×8	4Ø6	0.25	23.3[a]	345	460
SCESC1	280×280	1200	H150×150×7×10	12Ø16	0.0	29.5[b]	296	350
SCESC2	280×280	1200	H150×150×7×10	12Ø16	0.0	28.1[b]	296	350
SCESC3	280×280	1200	H150×150×7×10	12Ø16	0.0	29.8[b]	296	350
SCESC4	280×280	1200	H150×75×5×7	12Ø16	0.0	28.1[b]	303	350
SCESC5	280×280	1200	H150×75×5×7	12Ø16	0.0	26.4[b]	303	350
SCESC6	280×280	1200	H150×75×5×7	12Ø16	0.0	29.8[b]	303	350

[a] Concrete cube strength
[b] Concrete cylinder strength
[c] Mean value of steel flange and steel web yield strengths

confined concrete zone is from the parabola of the highly confined concrete zone to the centrelines of the transverse reinforcement as illustrated in Figure 3.10. The confining factor is expressed as (Mirza and Skrabek, 1991):

$$K = -1.254 + 2.254\sqrt{1 + \frac{7.94f_l'}{f_{c0}'}} - \frac{2f_l'}{f_{c0}'},$$

in which f_{c0}' is concrete unconfined concrete strength and f_l' is effective lateral confining stress depending on the volumetric ratio of lateral reinforcement, the configuration of the lateral and longitudinal reinforcement, and the area of the effectively confined concrete core. The confinement factor for highly confined concrete varied from 1.10 to 1.97 and for partially confined concrete varied from 1.08 to 1.50 depending on the spacing of the stirrups, as given by Chen and Lin (2006). The concrete outside the ties is not confined. The effect of residual stresses in structural steel is included and the initial imperfection is taken equal to L/1000 in which L is the effective length.

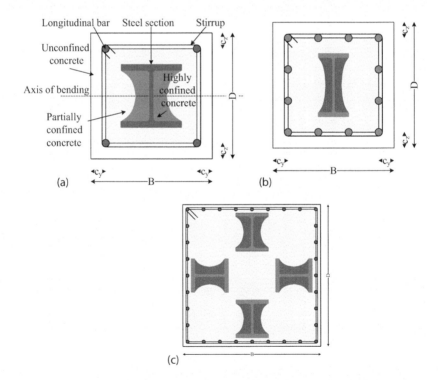

Figure 3.10 Specimen dimension and regions for unconfined, partially confined and highly confined concrete. (a) Specimen CESC1–CESC10. (b) Specimen SCESC1–SCESC6. (c) Specimens MCSRS1–SCSRS6

For all numerical simulations, the modified concrete stress-strain model proposed by Kent and Park (1971) in compression is adopted. For concrete in tension, linear stress-strain relationship up to tensile strength and linear tensile softening with fracture energy 0.12 N/mm are assumed. The stress-strain relationships of structural steel recommended by Eurocode 3 (2005a) and reinforcing bar recommended by Eurocode 2 (2005a) are adopted. All test specimens are modelled with the FE model presented earlier using six elements. Apart from specimens MCSRS1–MCSRS6, full shear interaction of concrete and steel components is assumed. For specimens MCSRS1–MCSRS6, shear connector stiffness at the extremities of the column is magnified to take into account the stiffeners at the column ends and shear connector strength is evaluated by using the Eurocode 4 (2005a) formulation. In Table 3.2, the predictions of the model are compared against test results. A good agreement between numerical and experimental results can be observed. Indeed, the mean value of the numerical-experimental load capacity ratio for 22 cases is very close to 1.0, and the corresponding

Table 3.2 Comparison between tests and finite element results

Specimen	λ	P_{Test} [kN]	P_{FE} [kN]	P_{FE}/P_{Test}
MCSRS1	20.78	17082	17697	1.04
MCSRS2	20.78	15325	16730	1.09
MCSRS3	20.78	14360	12282	0.86
MCSRS4	20.78	13231	13219	1.00
MCSRS5	20.78	12041	11322	0.94
MCSRS6	20.78	12759	12045	0.94
CESC1	30.12	654	641	0.98
CESC2	30.12	558	553	0.99
CESC3	30.12	962	813	0.85
CESC4	30.12	949	924	0.97
CESC5	45.18	900	822	0.91
CESC6	45.18	813	764	0.94
CESC7	45.18	704	646	0.92
CESC8	20.78	740	600	0.81
CESC9	51.96	504	493	0.98
CESC10	77.94	412	378	0.92
SCESC1	14.85	4220	4261	1.01
SCESC2	14.85	4228	4239	1.00
SCESC3	14.85	4399	4641	1.06
SCESC4	14.85	3788	3606	0.95
SCESC5	14.85	3683	3615	0.98
SCESC6	14.85	3893	3873	0.99
Mean	-	-	-	0.96
Cov	-	-	-	0.07

standard deviation is only 7%. Furthermore, it is worth mentioning that, in most cases, the FE model predictions are on the safe side.

3.4.3 Simplified second-order analysis of slender elements based on a moment magnification factor

In this section, the developed FE model which was successfully validated above is used to conduct parametric studies in order to assess the applicability of simplified methods of Eurocode 2 (2005a) and Eurocode 4 (2005a) for the hybrid columns, as well as to develop a new equation of effective flexural stiffness of hybrid columns. To do so, slender hybrid columns with different types of cross-sections are modelled using the present FE model.

3.4.3.1 Component considered for the parametrical studies

The first and second sections are hybrid columns with three encased steel profiles with the same European section, HEB120, where the bending axis is perpendicular to its strong and weak axis, Figure 3.11 and Figure 3.12, respectively. The third section is a mega-column, as seen in Figure 3.13, with four encased HD400x1086 steel profiles. The bending axis perpendicular

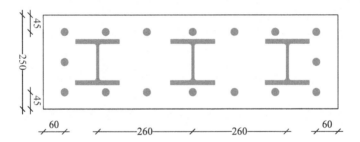

Figure 3.11 Cross-section HSRCC1

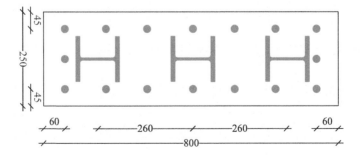

Figure 3.12 Cross-section HSRCC2

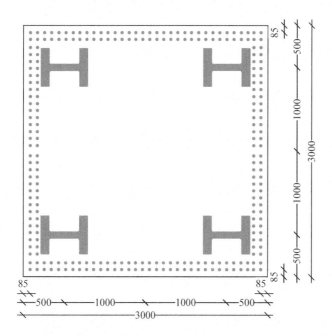

Figure 3.13 Cross-section HSRCC3

to the weak-bending axis of steel profiles is studied, and for the reason of symmetry, only a half-section is considered. For sections HSRCC1 and HSRCC2, the diameter of the reinforcement rebar is 20 mm and 32 mm for section HSRCC3. For all cases in this assessment, the limit of elasticity for the steel profile is restricted to 355 MPa, and for the reinforcement bar it is 500 MPa. Three classes of concrete strength, C35/45, C60/75 and C90/105, are considered. Sections HSRCC4 and HSRCC5 (Figure 3.14 and Figure 3.15) with a significant value of steel contribution ratio δ, evaluated according to Eurocode 4 (2005a) formulation, are studied for concrete class C35/45. Section HSRCC4 is not very realistic. It is studied here only for maximizing the ratio δ.

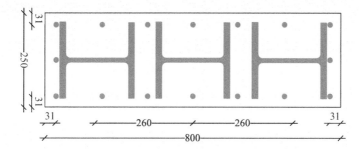

Figure 3.14 Cross-section HSRCC4

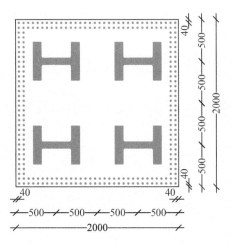

Figure 3.15 Cross-section HSRCC5

In high-rise buildings, there is a significant amount of long-term loads (approximately 75% of total loads). Therefore, the effect of sustained loads has to be considered. In this work, the effective creep ratio φ_{ef} is taken equal to 1.5. Therefore, the concrete stress-strain curve is modified following the Eurocode 2 (2005a) recommendation.

For columns subjected to axial compression and bending moment, three different column slenderness λ (low, medium and high slenderness) are considered for each cross-section configuration with or without taking into account the creep effect ($\varphi_{ef}=0$ or $\varphi_{ef}=1.50$). From the value of column slenderness and geometry of the cross-section, the column length can be deduced. For columns subjected to compressive load only, the whole range of possible column slenderness is covered. The parametrical study of 1140 data sets is summarized in Table 3.3.

In this study, bending is considered to take place about the strong axis (bent in symmetrical single curvature ($r_m=1$), in single curvature ($r_m=0$) and double curvature ($r_m=-1$)). This situation corresponds to the case where the extreme load is produced by wind or seismic load in that direction, and the motion of the column is restrained in the other direction.

3.4.3.2 Understanding of the physical behaviour of hybrid columns

Primarily, some effects on the nonlinear behaviour of hybrid columns with multiple embedded steel profiles are highlighted before presenting the proposed equations for the correction factors used in calculating the effective flexural stiffness of hybrid columns.

Table 3.3 Summary of case studies

Cross-section	HSRCC1 - HSRCC5
Concrete strength	C35/45; C60/75; C90/105
Rebar yield strength f_{yk}	500 MPa
Steel profile yield strength f_a	355 MPa
Column slenderness λ	20–185
Load eccentricity e/h	0.0–3.0
Steel profile contribution ratio δ	0.2–0.62
Creep effective ratio φ_{ef}	0; 0.15
End moment ratio r_m	–1; 0; 1

3.4.3.2.1 Effect of sustained loads

For a concrete column subjected to sustained loads, the deformation of the column tends to increase with time due to concrete creep. As a result, under axial force and bending moment produced by sustained loads, the column may lose its stability with time. It is the reason why concrete creep should be taken into account for slender column design. The most representative concrete creep model would be the time-dependent stress-strain relationship of concrete. With low-stress values, the creep law may be considered linear, and the critical buckling time of the column is infinite, whereas, for a non-linear creep law, the critical buckling time can be finite, as has often been observed in tests of concrete columns. The simple way of considering creep in the finite element analysis is a one-time-step calculation by modifying the concrete stress-strain curve. A comparison between the simplified and accurate models of concrete creep was performed by Westerberg (2002). The main conclusion was that the simplified approach is sufficiently accurate for practical purposes. Therefore, this simplification can be adopted in the finite element analysis of hybrid compression components.

Figure 3.16 shows the interaction curve obtained from FE analysis for section HSRCC2 with steel grade S355 and concrete class C35/45 for φ_{ef}=0 and φ_{ef}=1.50. $N_{pl,Rd0}$ and $M_{pl,Rd0}$ correspond to the plastic design normal force and moment of the cross-section without creep effect, respectively. It can be seen that the plastic design moment of the cross-section with φ_{ef}=1.50 is greater than that φ_{ef}=0. This is due to the large ductility of concrete, which allows the compression steel to yield before concrete crushes. Besides, we can see that the bearing capacity of the column with high slenderness is reduced when the concrete creep is considered.

3.4.3.2.2 Effect of residual stresses in structural steel

Residual stresses are a major imperfection of steel structural components because of their significant effects on buckling behaviour. Their distribution

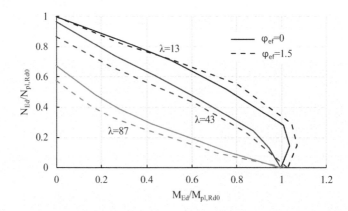

Figure 3.16 Effect of concrete creep on the interaction curve of hybrid columns

is shown in Figure 3.9. The interaction curve for sections HSRCC2 and HSRCC3 with steel grade S235 and concrete class C35/45 for φ_{ef}=1.50 is illustrated in Figure 3.17. The interaction curve with "o" marker is obtained when the residual stresses of the steel profile are not considered, while the solid line (—) is the one when the residual stresses are taken into account. It can be seen that the residual stresses have a low effect on the behaviour of hybrid columns and they can be ignored.

3.4.3.3 Assessment of Eurocode 2 (2005) moment magnification method

In the present section, the applicability of the Eurocode 2 (2005a) version of the moment magnification method to hybrid columns is assessed firstly by comparing its predictions against FE analysis results for hybrid columns with cross-section HSRCC1 (see Figure 3.11). S235 steel grade and C60 concrete class are considered. The effect of creep is taken into account by considering φ_{ef} equal to 1.5. It can be seen from Figure 3.18a that if the hybrid column is subjected to pure compression, the moment magnification method of Eurocode 2 (2005a) gives unsafe results for low-to-moderate column slenderness, whereas the method provides conservative results for high column slenderness. For columns subjected to single curvature bending and regardless of the load eccentricity (see Figure 3.18b and Figure 3.18c), Eurocode 2 (2005a) formulation overestimates the ultimate load for low-to-moderate column slenderness. The same conclusion applies to columns bent in double curvature under an antisymmetric bending moment (see Figure 3.18d) except for very high load eccentricities (close to pure bending). For high column slenderness, Eurocode 2 (2005a) method gives safe results except for columns bent in single curvature under a large bending moment.

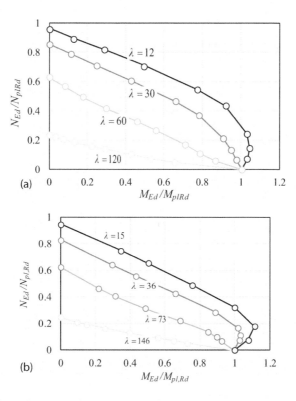

Figure 3.17 Effects of residual stresses on the interaction curve of hybrid columns. (a) Interaction curve for HSRCC2. (b) Interaction curve for HSRCC3

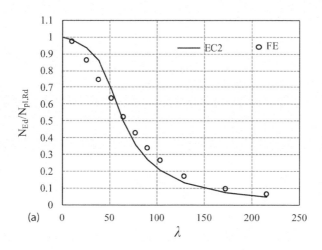

Figure 3.18a Comparison of the simplified method of EC2 and FEA. (a) Buckling curve

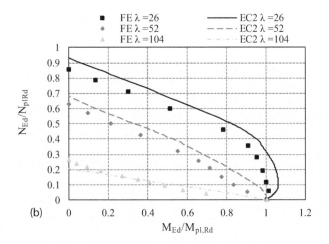

Figure 3.18b Single curvature bending with equal end moment

Since this simplified method is based on the effective flexural stiffness of the column $(EI)_{ef,II}$, it can be concluded that the equation for the effective flexural stiffness proposed by Eurocode 2 (2005a) cannot be applied in a straightforward fashion to hybrid column design. This effective flexural stiffness should be modified by adjusting the factor K_c, which depends on the relative slenderness of the column so that it becomes applicable to hybrid columns. Moreover, the factor K_s, which is applied to the stiffness, can also be modified in order to account for the yielding of the steel section.

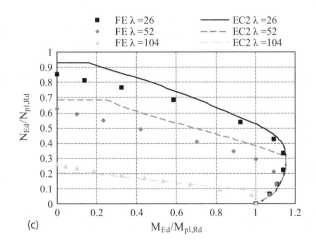

Figure 3.18c Single curvature bending with one end moment equal to zero

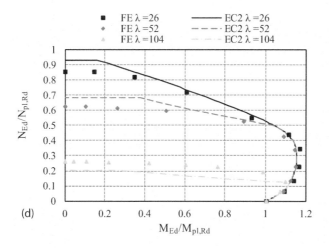

Figure 3.18d Double symmetrical curvature bending

3.4.3.4 Assessment of the Eurocode 4 (2005) variant of the moment magnification method

In this section, we pursue our study by an assessment of the performance of the Eurocode 4 (2005a) version of the moment magnification method when applied to hybrid columns. Again, a comparison of the predictions of the Eurocode 4 (2005a) method against FE analysis results for a hybrid column with cross-section HSRCC1 (see Figure 3.12) is first carried out. The steel grade and concrete class as well as effective creep ratio are the same as the previous case (S235, C60 and φ_{ef}=1.5). Quite surprisingly, the Eurocode 4 (2005a) version of the moment magnification seems to perform less well. Indeed, for a hybrid column subjected to pure compression (see Figure 3.19a) where the ultimate load of the column is characterized by the resistance in axial compression, the simplified method of Eurocode 4 (2005a) gives safe results regardless of column slenderness. Apart from that, this method gives unsafe results for a large number of cases.

For low load eccentricity, the ultimate load given by the Eurocode 4 (2005a) formulation is safe regardless of column slenderness (see Figure 3.19b to Figure 3.19d). For moderate load eccentricity, the Eurocode 4 (2005a) method always overestimates the ultimate load. Under large bending moment, the Eurocode 4 (2005a) method gives safe results, particularly for columns under symmetric single curvature bending in the zone, nearly pure bending. The conservative nature of the results can be attributed to the equivalent moment factor β, which, in the present case, is equal to 1.1.

Since this moment magnification method is based on the effective flexural stiffness of the column $(EI)_{ef,II}$, it can also be concluded that Eurocode 4 (2005a) proposes an equation for effective flexural stiffness that cannot be

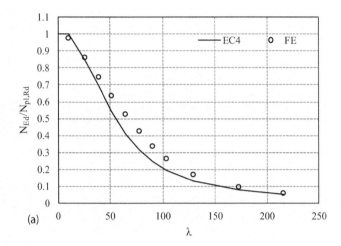

Figure 3.19a Comparison of the simplified method of EC4 and FEA. (a) Buckling curve

applied in a straightforward fashion to hybrid column design. This effective flexural stiffness (see Equation (3.2)) should be modified by reformulating the factor $K_{e,II}$ as well as K_0. These factors should be minimized to reduce the value of the effective flexural stiffness and as a result, the ultimate load would be decreased.

3.4.3.5 Synthesis of the parametric study

The ultimate load for isolated hybrid columns with five different cross-sections (HSRCC1 to HSRCC5) has been evaluated using both the finite element

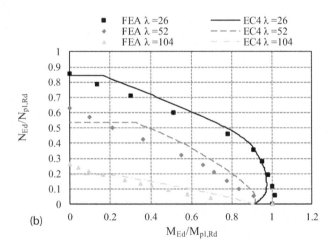

Figure 3.19b Single curvature bending with equal end moment

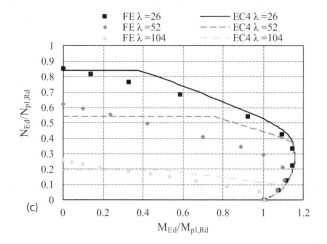

Figure 3.19c Single curvature bending with one end moment equal to zero

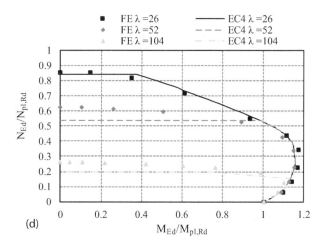

Figure 3.19d Double symmetrical curvature bending

model and the simplified method based on the moment magnification factor proposed in Eurocode 2 (2005a) and Eurocode 4 (2005c). The accuracy of the simplified method should be evaluated according to Appendix D of EN 1990 (2003). The application of this evaluation method is rather straightforward provided that a large number of ultimate loads are available with the magnitude of the latter being influenced by a single parameter. It is much more complex to apply this method for components subjected to axial load and bending moment where additionally, a large number of key parameters have to be taken into account. Because of these difficulties, an exact implementation of EN 1990-Appendix D (2003) cannot be rigorously followed while assessing the moment magnification method of Eurocode 2 (2005a) and Eurocode 4 (2005c). To evaluate the formulation of Eurocode 3 (2005b) for steel beam-column components, the ratio of the experimental or numerical reference failure load to the corresponding theoretical load is used in Boissonnad et al. (2006). Similarly, the ratio between the first-order bending moments obtained with numerical simulation $(M_1)_{FE}$ and the ones obtained via the simplified method $(M_1)_{SM}$ were used in Westerberg (2008) to calibrate the simplified methods of Eurocode 2 (2005a). However, this procedure is not appropriate in case columns subjected to axial load only which leads this ratio to infinity. To overcome this inadequacy, the ratio R expressed in (3.7) has been selected by Bonnet et al. (2011) as a reference value to evaluate the accuracy of their proposal. This ratio is also adopted in our investigation.

$$R = \frac{R_{FE}}{R_{SM}} \tag{3.7}$$

where:

$$R_{FE} = \sqrt{\left(\frac{N_{FE}}{N_{pl,Rd}}\right)^2 + \left(\frac{M_{FE}}{M_{pl,Rd}}\right)^2} \text{ and } R_{SM} = \sqrt{\left(\frac{N_{SM}}{N_{pl,Rd}}\right)^2 + \left(\frac{M_{SM}}{M_{pl,Rd}}\right)^2}.$$

Table 3.4 gives a summary of the results obtained with both the Eurocode 2 and Eurocode 4 versions of the moment magnification approach. which are compared against the results of FE analysis. In order to evaluate the contribution of the various parameters governing the ultimate load, the R ratio is first computed for all the considered cases (1140 data sets). The value of the R ratio is given as a function of each key variable: column slenderness λ, eccentricity e/h, steel contribution ratio δ, reinforcement ratio ρ, concrete characteristic strength f_{ck} and effective creep ratio φ_{ef}. For every value of each parameter, all corresponding values of R are given as discrete points.

To analyze the relative performance of the Eurocode 2 and Eurocode 4 variants of the moment magnification method, the graphs for the R ratio computed for each method for a given parameter are put as a pair. Regarding the contribution of the column slenderness on the variant of the method, two different graphs are provided. The first graph is for columns subjected to pure compression and the other one is for columns subjected to combined compression and bending. The statistical distribution of R is represented along with its mean value r and the interval $(r+s$ and $r-s)$ where s is the standard deviation. Both simplified methods show a rather wide discrepancy compared to FE analysis results. The most significant parameters are the slenderness of the column, the steel section contribution to the cross-section strength under pure compression δ as well as the geometrical reinforcement ratio ρ. Table 3.4 shows that for columns subjected to an axial load only (zero eccentricity), both simplified methods give unsafe results for low column slenderness. If the latter is moderate, the predictions of the Eurocode 2 moment magnification method are unsafe while Eurocode 4 one gives conservative results. Nevertheless, the Eurocode 2 method provides reasonable results compared to the Eurocode 4 method for high column slenderness. For columns subjected to combined compression and bending moment, both codes provide unsafe results in most cases. In particular, the interaction curve given by the Eurocode 2 moment magnification method without considering the creep effect ($\varphi_{ef}=0$) is close to FE analysis results. However, Eurocode 2 becomes unconservative if creep is considered ($\varphi_{ef}=1.5$). Considering all cases, it was found that the mean value r and the standard deviation s are, respectively, equal to 0.996 and 0.104 for Eurocode 2 simplified method, and 1.010 and 0.112 for Eurocode 4 simplified method. The percentage of R below 0.97 is 41.84% and 34.86% for the Eurocode 2 and Eurocode 4 simplified methods, respectively. As a general conclusion, it can be pointed out that mean estimations of both design codes seem to be correct but that their shortcomings lead to a large scatter of the results.

3.4.3.6 Development of the hybrid-specific variant of the moment magnification method

The parametric study with 1140 data sets presented previously shows that the Eurocode 2 and Eurocode 4 simplified methods lead to unsafe results in half of the case studies. It means that the proposed effective flexural stiffness $(EI)_{eff,II}$ of Eurocode 2 and Eurocode 4 are not appropriate for slender hybrid column design. In this study, based on a new parametric study with 2960 cases including different strengths of steel profile, new equations for the correction factors (for the determination of effective flexural stiffness $(EI)_{eff,II}$ expressed in Equation (3.5)) are proposed. It can be seen that the equations of the correction factors K_c is the one recommended in Eurocode 2. Because there is no steel profile in the reinforced concrete section, the correction factor K_{sa} on the flexural stiffness of the steel profile does not exist in Eurocode 2. If one compares these correction factors to those in Eurocode 4, they are very different. In fact, due to time-dependent concrete strains as shown in Section 3.4.3(b), longitudinal steel compression strain can be higher than its yield strains. This implies that its modulus that collaborates in the effective flexural stiffness $(EI)_{eff,II}$ of the column should not be the elastic modulus but the secant modulus that will depend on the creep of concrete. Moreover, for slender columns, the yielding of the section could not be reached due to instability. Hence, the secant modulus of steel should be in function of the creep coefficient φ_{ef} and of the geometric slenderness λ. For higher values of the creep coefficient, the value of the secant modulus of steel will be lower; and it will be higher for higher values of slenderness. Therefore, in addition to previous cases for the assessment of the moment magnification method of Eurocode 2 and Eurocode 4, analyses with two

Table 3.4 Synthesis of parametric studies

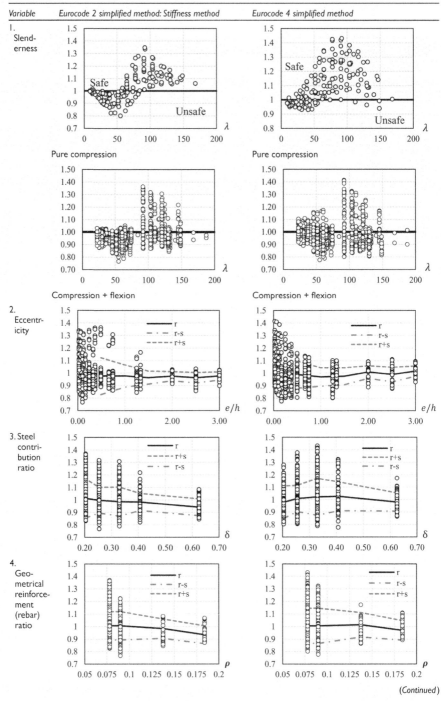

(Continued)

Table 3.4 (Continued) Synthesis of parametric studies

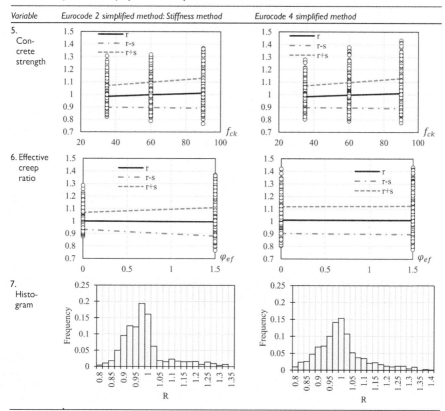

additional yielding stresses of steel profiles (f_a=235 MPa and f_a=460 MPa) are carried out. The objective is to study the effect of the yielding of steel profiles. As a result, compared to the equations in Eurocode 4, the correction factor K_{sa} is modified to take into account the effect of yielding of the steel section. This factor is calibrated from the results of a parametric study with 2960 parameter sets (cross-sections, material strength, column slenderness and creep coefficient) performed by using the developed FE model.

The procedure for determining the equation of K_{sa} is as follows. Let us consider a slender hybrid column with initial imperfection w_0 subjected to axial loads and uniaxial bending, bent in symmetrical single curvature (r_m=1), the ultimate first-order bending moment $M_{Ed,1}$ can be obtained for a particular axial load N_{Ed} if a nonlinear FE analysis is performed. Likewise, it is also possible to compute the ultimate bending moment $M_{pl,N,Rd}$ of the cross-section of the column for the same axial force. By equating the second-order bending moment determined by using the moment magnification method to the ultimate bending moment of the cross-section of the column, the moment magnification factor k can be obtained. Finally, by making use of the critical buckling load formulation and the proposed form of an effective flexural stiffness equation (adopting the correction factor on concrete stiffness K_c, such as the one in Eurocode 2), the correction factor K_{sa} can be derived. It is worth mentioning that for a low percentage of steel reinforcement in cross-section, the hybrid column can be considered a reinforced concrete one, since residual stresses in steel sections have no influence (negligible) on the column behaviour. Hence, the correction factors proposed in Eurocode 2 can be adopted for a low percentage of steel reinforcement.

From an extensive numerical parametric study with 2960 data sets, the following equation of K_{sa} has been proposed:

$$K_{sa} = \begin{cases} 1, & if \ \rho_{sa} = \dfrac{A_s + A_a}{A_c} \le 0.04 \\[2em] \dfrac{0.450\lambda^{0.147}}{1 + 1.433\varphi_{ef}exp(-0.027\lambda)}, & if \ \rho_{sa} = \dfrac{A_s + A_a}{A_c} > 0.04 \end{cases}$$

3.5 A DEDICATED SOFTWARE FOR HYBRID COMPONENTS: *HBCOL*

HBCOL is a software dedicated to implementing the simplified method introduced in Section 3.3.4 to predict accurately the ultimate loads of hybrid columns reinforced by several embedded steel profiles. This software is also capable of performing the nonlinear analysis of hybrid columns subjected to combined axial load and uniaxial bending moment by adopting the finite element model presented in Section 3.4.2.

3.5.1 Limitations of HBCOL

HBCOL is subject to the following limitations:

- The accuracy of the finite element model is dependent on the cinematics assumptions, the constitutive model for materials and convergence criteria used in finite element codes.
- Full interaction of shear connection at the steel-concrete interface is assumed in the finite element model. In other words, interface slips are not considered.
- Only one mechanical property is used for each structural component (steel section, reinforcing bars, concrete).
- The analysis can be performed for hybrid columns with three embedded steel sections at most.
- All steel sections have to have the same orientation of the cross-section, for example, I-orientation or H-orientation.

- Steel sections defined by fibres (round section) can be used only in the simplified method.
- The compatibility strain method (see Section 3.3.6.1) is the only approach used to determine the interaction curve of the full plastic cross-section.

3.5.2 Data interface

The data interface (Figure 3.20) allows defining:

- The column dimensions, including the eccentricity (used for nonlinear analysis) of the normal forces at both ends.
- The distribution of the reinforcement bars.
- The number and position of the steel profiles in the section.
- The steel section defined by fibres (only taken into account in the simplified analysis).

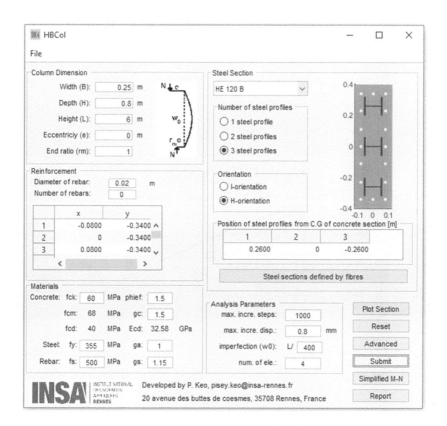

Figure 3.20 Data interface of HBCOL

- The material laws of steel and concrete, including the safety coefficients adopted.
- The parameters for analysis, including the amplitude of the sinusoidal imperfection.

3.5.3 Results

Once data are defined, the button "Simplified M-N" plots the interaction curve given by the simplified design method (see Section 3.3.6). Two interaction curves are given (see Figure 3.21):

- The plastic cross-section interaction curve, which gives the sectional resistance.
- The column interaction curve, which gives the resistance of the column taking into account the second-order effects.

The button "Submit" runs the nonlinear analysis for the specific M-N case defined by the eccentricity ratio. This nonlinear analysis takes into account the residual stresses in the steel profiles. It gives as a result the force-displacement curve (axial force vs axial displacement, see Figure 3.22).

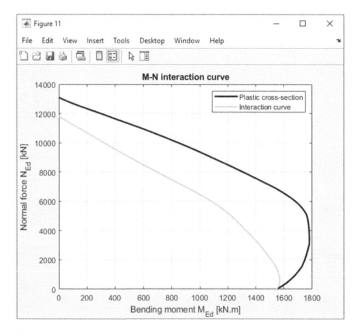

Figure 3.21 Interaction curves given by simplified method

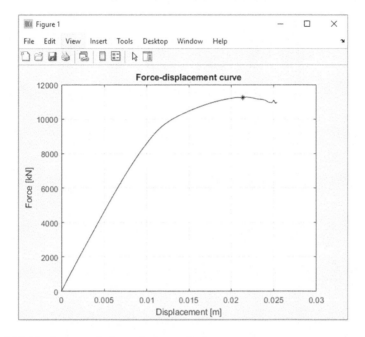

Figure 3.22 Nonlinear analysis result: force-displacement curve

3.6 DESIGN EXAMPLE

A mega-column of **31 m** buckling length with four embedded steel sections HD400x1086 positioned at the four corners of the cross-section is considered (see Figure 3.23). Reinforcing bars of 32 mm diameter are positioned at the peripheral side of the column in two layers with a total number of 256. The concrete cover is 85 mm and the spacing between the two layers is 83 mm.

Material properties have been chosen as follows:

- Structural steel: grade S355, $f_y = f_{yd} = 355$ N/mm², $E_a = 210$ kN/mm².
- Concrete: C35/45; $f_{ck} = 35$ N/mm², $f_{cd} = 35/1.5 = 23.3$ N/mm², $E_{cm} = 34$ kN/mm², $\varphi_{ef} = 1.5$.
- Reinforcement: grade BA S500, $f_{sk} = 500$ N/mm², $f_{sd} = 500/1.15 = 435$ N/mm².

Characteristic of the cross-section:

- Concrete: $A_c = 8.2406$ m², $I_c = 5.9134$ m⁴
- Rebars: $A_s = 0.2059$ m², $I_s = 0.2593$ m⁴
- Steel shape: $A_a = 0.5535$ m², $I_a = 0.5773$ m⁴

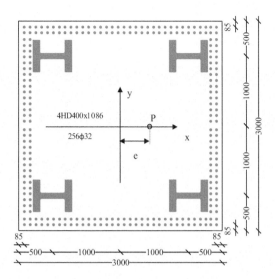

Figure 3.23 Column cross-section (dimension in mm)

Note: the moment of inertia is calculated at the centre of the area of the gross concrete section.

Considering a geometric imperfection taken as $w_0=L/400$, one can determine the first-order bending moment $M_{Ed,1}$ of symmetrical single curvature bending with the applied moment and normal force at the extremity (N_{Ed}, M_{Ed}) as:

$$M_{Ed1} = M_{Ed} + N_{Ed} \times w_0 = N_{Ed}\left(e + w_0\right)$$

in which e is the eccentricity.

Considering the design value of the compressive normal force $N_{Ed}=301851$ kN and the eccentricity of **0.3 m**, we obtain the first-order bending moment M_{Ed1}:

$$M_{Ed1} = 301851 \times \left(0.3 + \frac{31}{400}\right) = 113949\,kN.m$$

3.6.1 Results of the nonlinear analysis

A nonlinear FE analysis can be performed using HBCOL software, considering a demi-section of the column. With an eccentricity of 0.3 m, the collapse is reached for a load multiplier equal to 1.75 for a couple of linear forces in mid-section, $N_{Ed} = 301\ 851$ kN and $M_{Ed1} = 113\ 949$ kN.m (the collapse load obtained from FEA is multiplied by 2 in order to consider the full cross-section of the column).

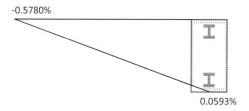

-0.5780%

0.0593%

Figure 3.24 Deformation distribution of the critical cross-section at the ultimate load

The deformation distribution of the critical cross-section at the ultimate load is shown Figure 3.24. It can be observed that the steel profile in compression is completely yielded.

Calculating the secant modulus of each fibre for each component and summing the secant flexural stiffness of each fibre within the whole cross-section, we obtain:

$$\left(EI\right)_{c,sec} = 13580 \times 2 = 27160 \,\text{MN.m}^2$$

$$\left(EI\right)_{s,sec} = 18929 \times 2 = 37857 \,\text{MN.m}^2$$

$$\left(EI\right)_{a,sec} = 41636 \times 2 = 83271 \,\text{MN.m}^2$$

We can compute the corresponding correction factors:

$$\left(EI\right)_{eff} = \left(EI\right)_{c,sec} + \left(EI\right)_{s,sec} + \left(EI\right)_{a,sec}$$

$$= \frac{\left(EI\right)_{c,sec}}{\left(E_{cd}I_c\right)}\left(E_{cd}I_c\right) + \frac{\left(EI\right)_{s,sec}}{\left(E_sI_s\right)}\left(E_sI_s\right) + \frac{\left(EI\right)_{a,sec}}{\left(E_aI_a\right)}\left(E_aI_a\right)$$

$$= \frac{27160}{34000 / 1.2 \times 5.9134}\left(E_{cd}I_c\right) + \frac{37857}{210000 \times 0.2593}\left(E_sI_s\right) + \frac{83271}{210000 \times 0.5773}\left(E_aI_a\right)$$

$$= 0.1620\left(E_{cd}I_c\right) + 0.6946\left(E_sI_s\right) + 0.6862\left(E_aI_a\right)$$

$$= \bar{K}_c\left(E_{cd}I_c\right) + \bar{K}_s\left(E_sI_s\right) + \bar{K}_a\left(E_aI_a\right)$$

Thus, the correction factors are:

$$\bar{K}_c = 0.1620; \quad \bar{K}_s = 0.6946; \quad \bar{K}_a = 0.6862$$

This shows clearly that, at collapse, the secant stiffness must be computed with values of K_s and K_a lower than 1.00.

In the following, we compare the simplified methods proposed in Eurocode 2, Eurocode 4 and the one presented in Section 3.3.6.

3.6.2 Application of the simplified method for slender hybrid column

Now the same column is computed following the simplified method, for the action at collapse in FE analysis.

Resistance of the cross-section:

The resistance of the cross-section of the hybrid column is determined using the compatibility strain method. We obtain then the interaction curve of resistance of cross-section as illustrated in Figure 3.25.

- The plastic resistance of the cross-section to compressive normal force is $N_{pl,Rd}$=478299 kN;
- The plastic resistance moment of the cross-section is $M_{pl,Rd}$=320501 kN.m.

The plastic resistance moment of the cross-section taking into account the compressive normal force N_{Ed}=301 851 kN is $M_{pl,N,Rd}$=199 397 kN.m

Contribution factors:

- Concrete:

$$\lambda = \frac{L}{\sqrt{\dfrac{I_g}{A_g}}} = \frac{31}{\sqrt{\dfrac{3 \times \dfrac{3^3}{12}}{3 \times 3}}} = 35.7957$$

$$k_1 = \sqrt{\frac{f_{ck}}{20}} = \sqrt{\frac{35}{20}} = 1.3229; \quad n = \frac{N_{Ed}}{N_{pl,Rd}} = \frac{301851}{478299} = 0.6311;$$

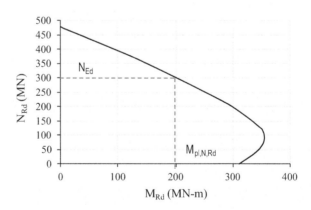

Figure 3.25 Resistance interaction curve of cross-section obtained from HBCol software using compatibility strain method

$$k_2 = n\frac{\lambda}{170} = 0.6311 \times \frac{35.7957}{170} = 0.1329 \le 0.2$$

$$K_c = \frac{k_1 k_2}{1 + \varphi_{ef}} = 1.3229 \times \frac{0.1329}{1 + 1.5} = 0.0703$$

- Rebar and steel shape:

$$K_{sa} = \frac{0.450 \times 35.7957^{0.147}}{1 + 1.433 \times 1.5 \times 1.5 \times \exp(-0.027 \times 35.7957)} = 0.4189$$

Nominal stiffness:

$$\begin{aligned}
(EI)_{eff,II} &= K_c E_{cd} I_c + K_{sa}(E_s I_s + E_a I_a) \\
&= \left[0.0703 \times \frac{34000}{1.2} \times 5.9134 + 0.4189 \times 210000 \times (0.2593 + 0.5773) \right] \\
&= 85373.2726 \, \text{MN.m}^2
\end{aligned}$$

Elastic critical normal force:

$$\begin{aligned}
N_{cr,eff} &= \frac{\pi^2}{L^2}(EI)_{eff,II} \\
&= \frac{3.14^2}{31^2} \times 85373.2726 = 876 \, \text{MN}
\end{aligned}$$

Magnification factor:

$$\beta = 0.6 + 0.4 r_m = 1 \ge 0.4$$

$$k = \frac{\beta}{1 - \dfrac{N_{Ed}}{N_{cr},eff}} = \frac{1}{1 - \dfrac{301.851}{876}} = 1.53$$

Second-order bending moment:

$$M_{Ed2} = k \times M_{Ed1} = 1.53 \times 113949 = 173856 \, \text{kN.m} < M_{pl,N,Rd} = 199397 \, \text{kN.m}$$

$$\frac{M_{Ed2}}{M_{pl,N,Rd}} = 0.87$$

Hence, the amplification factor estimates quite correctly the nonlinear moment $M_{Ed,2}$, with a difference of 13%. However, for this specific case, this is unconservative compared to FEA. Using the simplified method, the collapse is reached for a normal force $N_{Ed,SM}$ equal to 315.513 MN for a given eccentricity 0.3 m, meaning an overestimation of the collapse load of

4.3% with a ratio $R = \dfrac{\sqrt{\left(\dfrac{N_{Ed,FE}}{N_{pl,Rd}}\right)^2 + \left(\dfrac{M_{Ed,1,FE}}{M_{pl,Rd}}\right)^2}}{\sqrt{\left(\dfrac{N_{Ed,SM}}{N_{pl,Rd}}\right)^2 + \left(\dfrac{M_{Ed,1,SM}}{M_{pl,Rd}}\right)^2}} = 0.957.$

Comparison of correction factors K_s and K_a of simplified method with their numerical value:

Comparing the correction factors to the ones in the proposed design formulation in determining the effective flexural stiffness with the applied axial load N_{Ed} equal to the ultimate load N_u=301851 kN, we obtain the following factors:

K_c=0.0703 and K_{sa}=0.4189

We can observe that the correction factor on the concrete stiffness that has been chosen following Eurocode 2 is significantly lower than the one obtained from the nonlinear analysis result. If we make a sum of the secant flexural stiffness of both steel components obtained from nonlinear analysis result we have:

$$\left(EI\right)_{s,sec} + \left(EI\right)_{s,sec} = 37857 + 83271 = 121128 \, \text{MN.m}^2$$

while using the proposed design formulation gives:

$$K_{sa}\left(E_s I_s + E_a I_a\right) = 0.4189 \times 210000 \times \left(0.2593 + 0.5773\right) = 73595 \, \text{MN.m}^2$$

The ratio of the two is: $\dfrac{K_{sa}\left(E_s I_s + E_a I_a\right)}{\left(EI\right)_{s,sec} + \left(EI\right)_{a,sec}} = 0.61$

Therefore, the reduction factor of the simplified method is lower than the numerical one. However, it leads to an overestimation of the capacities of the column in this case (as $M_{Ed2}/M_{pl,N,Rd}$=0.87). In fact, K_c and K_{sa} are calibrated for the moment magnification method where the criteria to be satisfied is that the second-order bending moment is not greater than the plastic bending resistance $(M_{pl,N,Rd})$ taking into account the applied normal force. It means that the failure of the columns is supposed to be a material failure (resistance of cross-section). However, at collapse, the full yielding is not reached. The collapse is induced by instability before reaching full bending plastic capacity. It is worth mentioning that this phenomenon is well known and happens even for steel components. It is handled in Eurocode 3 (2005a) in the formulation of the French-Belgian approach through the coefficient C_{yy}, see (Boissonnade et al., 2004). It is also worth mentioning that this coefficient C_{yy} has been calibrated numerically, as this premature collapse is hard to grasp.

3.6.3 Application of Eurocode 2

If we adopt $K_s = K_a = 1$, the elastic critical normal force is then equal to $N_{cr,eff} = 1923$ MNm, which leads to the moment magnification factor k equal to 1.18. Multiplying this factor to the first-order bending moment gives the second-order bending moment $M_{Ed2} = 135400$ kNm. Thus,

$$\frac{M_{Ed2}}{M_{pl,N,Rd}} = 0.68$$

Consequently, the amplification is underestimated with a factor of 32 %. Using the simplified method of Eurocode 2 (2005a), the collapse is reached for a normal force equal to 342.791 MN for a given eccentricity 0.3 m, leading to an overestimation of the collapse load by 14 %.

3.6.4 Application of Eurocode 4

Applying the equations of correction factors proposed in Eurocode 4 (2005a), we have:

$$\left(EI\right)_{eff,ll} = 0.45\,E_{cm}I_c + 0.9\left(E_sI_s + E_aI_a\right) = 248592.42\,\text{MN.m}^2$$

which gives $N_{cr,eff} = 2553$ MN.

The moment magnification factor k is then equal to 1.13. Therefore, the second-order bending moment is $M_{Ed,2} = 129227$ kNm. Thus,

$$\frac{M_{Ed2}}{M_{pl,N,Rd}} = 0.65$$

The difference between the numerical results and those obtained by Eurocode 4 in the estimation of the collapse load is larger; it is about 35 %. The collapse is reached for a normal force equal to 345.939 MN for a given eccentricity of 0.3 m, meaning an overestimation of the collapse load of 15 %.

3.6.5 Summary of the results at collapse level

The results of applying the different methods are summarized in the table below at the collapse level:

	FEA	Eurocode 2	Eurocode 4	Simplified method
$N_{cr,eff}$ (kN)	1,522,000	1,941,725	2,553,079	882,284
k (theoretical)	1.75	1.21	1.13	1.56
N_{Ed} at collapse (kN)	301,850	342,791	345,939	315,513
$M_{Ed,I}$ (kNm)	114,000	129,404	130,592	119,106
$M_{pl,N,Rd}$ (kN)	199,396	157,148	151,078	185,398
N_{Ed}/N_{FEA}	1.00	1.14	1.15	1.05

This is a single example but illustrates well the reasons that have led to modify the factors K_s and K_a: adopting K_s and K_a equal to 1.00 as in Eurocode 2 (2005a) and to 0.9 as in Eurocode 4 (2005a) leads to overestimating the critical normal force. This example illustrates as well the over-amplification of the simplified methods produced by an underestimation of the critical normal force.

REFERENCES

Al-Shahari, A., Hunaiti, Y., and Ghazaleh, B. (2003) Behavior of lightweight aggregate concrete-encased composite columns. *Steel & Composite Structures*, 3(2), 97–110.

Bergmann, R., and Hanswille, G. (2006) New design method for composite columns including high strength steel. In: R. T. Leon & J. Lange, eds. *Composite construction in steel and concrete V*, ASCE, Mpumalanga, 381–389.

Bogdan, T., Chrzanowski, M., and Odenbreit, C. (2019) Mega columns with several reinforced steel profiles – Experimental and numerical investigations. *Structures*, 21, 3–21.

Bogdan, T., Degée, H., Plumier, A. and Campian, C. (2012) A simple computational tool for the verification of concrete walls reinforced by embedded steel profiles. In: SPES, *15th world conference on earthquake engineering*, Sociedade Portuguesa de Engenharia Sismica (SPES), Lisbon, Portugal, 1–10.

Boissonnade, N., Greiner, R., and Jaspart, J. P. (2006) *Rules for member stability in EN 1993-1-1. Background documentation and design guidelines*. ECCS Technical Committee 8-Stability, Brussels.

Boissonnade, N., Jaspart, J.-P., Muzeau, J.-P., and Villette, M. (2004) New interaction formulae for beam-columns in Eurocode 3: The French-Belgian approach. *Journal of Constructional Steel Research*, 60(3), 421–431.

Bonet, J., Romero, M., and Miguel, P. (2011) Effective flexural stiffness of slender reinforced concrete columns under axial forces and biaxial bending. *Engineering Structures*, 33(3), 881–893.

British Standard Institution. (2003) *BS EN 1990:2003. Eurocode: Basic of structural design*. BSI, London.

British Standard Institution. (2005a) *BS EN 1992-1-1:2005. Eurocode 2: Design of concrete strucutres. Part 1–1: General rules and rules for buildings*. BSI, London.

British Standard Institution. (2005b) *BS EN 1993-1-1:2005. Eurocode 3: Design of steel structures. Part 1–1: General rules and rules for buildings*. BSI, London.

British Standard Institution. (2005c) *BS EN 1994-1-1:2005. Eurocode 4: Design of composite steel and concrete structures. Part 1–1: General rules and rules for buildings*. BSI, London.

Chen, C.-C., and Lin, N.-J. (2006) Analytical model for predicting axial capacity and behavior of concrete encased steel composite stub columns. *Journal of Constructional Steel Research*, 62(5), 424–433.

Dan, D., Fabian, A., and Stoian, V. (2011) Theoretical and experimental study on composite steel-concrete shear walls with vertical steel encased profiles. *Journal of Constructional Steel Research*, 67(5), 800–813.

Kent, D. C., and Park, R. (1971) Flexural members with confined concrete. *Journal of the Structural Division*, 97(7), 1969–1990.

Keo, P., Nguyen, Q.-H., Somja, H., and Hjiaj, M. (2015a) Geometrically nonlinear analysis of hybrid beam–column with several encased steel profiles in partial interaction. *Engineering Structures*, 100, 66–78.

Keo, P., Somja, H., Nguyen, Q.-H., and Hjiaj, M. (2015b) Simp 110 lified design method for slender hybrid columns. *Journal of Constructional Steel Research*, 110, 101–120.

Mirza, S., and Skrabek, B. (1991) Reliability of short composite beam-column strength interaction. *Journal of Structural Engineering*, 117(8), 2320–2339.

Morino, S., Matsui, C., and Watanabe, H. (1985) Strength of biaxially loaded SRC columns. In: Charles W. Roeder, ed. *Composite and mixed construction: Proceedings of the U.S./Japan joint seminar*. ASCE, Washington, 185–194.

Westerberg, B. (2002) *Second order effects: Background to Chapters 5.8, 5.9 and Annex H in EN 1992-1-1*. KTH (Royal Institute of Technology in Stockholm), Stockholm.

Westerberg, B. (2008) *Time-dependent effects in the analysis and design of slender concrete compression members*. Civil and architectural engineering. Division of concrete structure, KTH, Stockholm.

Zhou, Y., Lu, X., and Dong, Y. (2010) Seismic behaviour of composite shear walls with multi-embedded steel sections. Part I: Experiment. *The Structural Design of Tall and Special Buildings*, 19(6), 618–636.

Chapter 4

Hybrid walls and columns

André Plumier

CONTENTS

DOI: 10.1201/9781003149811-4

4.1 INTRODUCTION

4.1.1 Scope of Chapter 4

This chapter explains the way to calculate the action effects and the resistance of walls and columns with several embedded steel profiles, which are referred to here as hybrid walls or hybrid columns. It also gives recommendations concerning their detailing. The types of sections envisaged are shown in Figure 4.1. They are characterized by embedded steel sections connected only by reinforced concrete, without connections by means of welded or bolted steel plates or sections, outside of those necessary at storey levels to connect beams or slabs to the columns or walls.

One design code, AISC 360-16, allows the design of composite sections with two or more encased sections and gives some limitations, but the way to design them is not detailed. Sections built from two or more encased steel shapes are also allowed by the Chinese Institute of Earthquake Engineering.

Design methods for composite columns with steel plate webs are largely present in publications, e.g. in Epackachi et al. (2015) or Varma et al. (2011) and they are covered in AISC341-16. They are not presented in this chapter, which focuses on symmetrical hybrid columns and wall sections of the types shown in Figure 4.1.

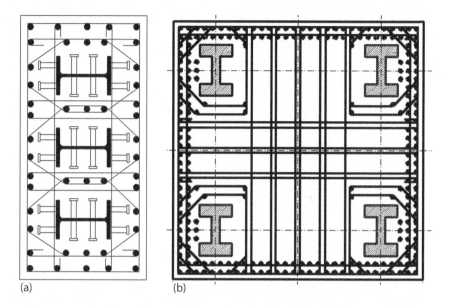

Figure 4.1 a, b Examples of hybrid sections. At left: a wall. At right: a column

The positive characteristics and design problems of walls and columns with several embedded steel profiles are first pointed out, followed by the definition of specific design issues. Then, the behaviour of hybrid components is studied. The subjects of global stiffness and instability of such components being already covered in Chapter 3, Chapter 4 focuses in Sections 4.2 to 4.5 on the internal action effects in walls and columns and, in Sections 4.6 and 4.7, on their resistance and ductility, envisaging successively the physics of the problems and the possible models. Chapter 4 aims to develop practical analytical design methods to be used in daily practice before any smarter numerical modelling. Numerical models of such walls are not discussed here. Load introduction and anchorages are envisaged in Section 4.8. Information and lessons from experimental activities are presented in Section 4.9, while Section 4.10 gives guidance on the details of longitudinal and transverse steel content. Design examples in Section 4.11 conclude Chapter 4.

4.1.2 Positive characteristics of walls and columns with several embedded steel profiles

Hybrid columns or walls, in which steel profiles are used as longitudinal steel reinforcements, present several advantages which can be summarized as follows.

More compact sections by avoiding reinforcement congestion.
Encased structural steel develops a composite action with concrete increasing the compression, bending and shear strength of walls. For a given action

effect, the necessary total cross-section area can be smaller if high-performance materials are used. Steel profiles have a high steel content without constraint, such as with respect to distances between reinforcing bars. This solves a technological issue that arises in boundary zones of walls or with face reinforcement of columns when high steel-to-concrete ratios are required. With bars, there may be reinforcement congestion in the wall or column boundary zones and difficulties with bar overlap or sleeve joints. The use of embedded compact steel sections thus permits a higher steel content in a given concrete section, which results in more free space at each storey. This is of interest since free space is what is sold to clients.

Shear stiffness and resistance to in-plane effects.

Steel profiles have a great shear stiffness in comparison to bars. They have also a significant shear resistance, which is not the case in classical reinforcement, and they are highly ductile in shear. Embedded steel profiles contribute to the shear stiffness of walls. However, the shear stresses in profiles affect their resistance to axial force, and this has to be considered in the calculation of the resistance of hybrid components to combined bending and axial force.

Resistance to out-of-plane effects in walls.

Failures of thin walls due to out-of-plane deformations were, for instance, largely observed after the 2010 earthquake in Chile. Embedded steel profiles can substantially increase the resistance of walls to out-of-plane bending or buckling, by providing a significant out-of-plane stiffness and strength. This contribution is particularly interesting in thin and slender walls with minimum amounts of classical reinforcement, in particular, if such walls are submitted to axial load ratios $N_{Ed}/A_c f_{cd}$ greater than 0.3. A layout in which the encased steel profiles provide their strong axis resistance, as shown in Figures 4.1a, 4.9 and 4.47a provides the larger possible contribution to resistance against out-of-plane bending.

Mitigation of sliding shear failure.

Steel profiles are good at mitigating sliding shear failure, a type of failure which takes place along weak planes in the concrete. Horizontal interfaces between concrete poured at different times generally exist as construction joints at the base of a wall and at every storey level, and are weak planes. They may be a cause of problems unless special measures are taken to increase bond and friction. As shown in Figure 4.2, encased steel profiles are much stronger dowels than classical reinforcement.

In an earthquake situation, a weak plane can be formed at the base of a wall by the accumulation of inelastic deformations in the vertical rebars as cracks open at the base of the wall that do not close fully at load reversals; base cracks facilitate sliding shear and conclude in a wall moving horizontally as a rigid body which submits the vertical reinforcement to alternate shear. Diagonal cracks generated by shear can also be weak planes in which

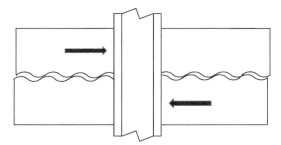

Figure 4.2 Encased profiles can prevent sliding shear and displacement in shear cracks

short segments of the vertical reinforcement crossing those planes behave as small vertical beams with two plastic hinges. Those reinforcements may fail because of a mix of damage due to plastic deformations and buckling of the bars. Encased steel profiles can prevent such failures.

Better resistance to tension and to combined shear and tension
Under earthquake or wind, outrigger columns or walls may be submitted to tension and concrete cracking, a key factor of stiffness degradation of the component. The lateral stiffness and shear strength of reinforced concrete walls or columns submitted to increasing axial tensile forces may decrease to approximately 25% of the values in the absence of axial tensile force. This favours shear-sliding or sliding failure, a phenomenon that can be better mitigated if steel sections are used as reinforcements because these can significantly increase the stiffness and strength of piers in tension.

Greater ductility in bending.
A compression chord essentially constituted by steel profiles is more ductile than a compressed chord made of concrete and of classical reinforcement, for the following reasons:

a) The inertia of the steel profiles is much greater than those of bars; their slenderness between lateral restraint by stirrups is thus much smaller than the one of bars. Steel profiles are normally chosen with compact sections in order to realize a high resistance to compression or bending. Compact sections are not prone to local buckling like classical bars and their ductility in compression can be exploited fully.

b) If rebars around a steel profile tend to bend and buckle, this reduces their axial stiffness. Given that the profiles have no such tendency, a redistribution of the action effect from the bars present in the section towards the profiles can take place, which delays the buckling of bars. This is possible if a large part of the longitudinal steel content in the section is constituted by steel profiles.

c) Under compression, steel is more ductile than concrete. With a large steel content and the absence of detrimental buckling of rebars, the ductility of walls or columns can be about two times greater than with a low steel content of classical bars. This can be assessed as follows. The yield strain ε_y of constructional steel and the strain ε_c in concrete at maximum stress are similar: $\varepsilon_y = 355/210000 = 0.00169$ for S355 steel; for concrete $\varepsilon_c = 0.002$, crushing of unconfined concrete corresponds to a strain $\varepsilon_{cu} = 0.0035$; the confinement of concrete by stirrups increases this failure strain by a factor about 1.25: $\varepsilon_{cc} = 0.004375$. As the ductility of steel profiles in compression is of the order of 10, the maximum strain for S355 steel is $\varepsilon_{y,max} = 0.0169$, approximately four times greater than the failure strain ε_{cc} of confined concrete. If the steel content due to steel profiles is high, the section in which concrete is crushed in compression does not lose much strength, and the remaining resistance is available with a ductility coefficient of about 4 if the concrete web between profiles is designed to remain intact.

Post-earthquake resistance.
After an earthquake, the structural system can remain stable thanks to the remaining resistance of the steel profiles. The steel profiles can keep a significant energy dissipation capacity after an earthquake, and this may be assessed by proper non-linear analysis. Repairs can be realized by replacement/injections of the cracked concrete in which the profiles are embedded.

Effectiveness in the building construction phase.
Over built-up caissons, which require complex positioning of multiple plates and long multi-pass weld length, steel profiles in hybrid columns only need longitudinal continuity welds. To gain in performance, the encased sections can be stocky rolled H sections with flange thickness up to 140 mm; these are available in high tensile strength steel produced by a quenching and self-tempering process, namely ASTM A913 Grade 345 and 450, or per ETA, European Technical Approval 10-156, grades Histar 355 and 460. These steels may be welded without preheating, they offer a high toughness and are characterized by an elongation greater than 20%. The welding procedures and the connection details of rolled H sections are described in the codes.

Easy connection with steel or composite beams and faster erection time.
Steel profiles encased in columns or walls can be easily connected with steel and composite steel-concrete floor beams that are often used in buildings. The connections are particularly simple in columns with more than two encased steel shapes because beams may be framing through the columns as shown in Figure 4.3, and a plate connecting two steel sections of the column can provide direct support to beam shear: two plates connecting

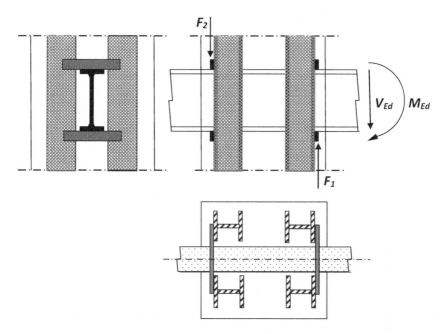

Figure 4.3 A beam connection to a hybrid column with four encased steel sections

two steel sections of the mega column provide two forces F_1 and F_2, which can equilibrate a beam moment M_{Ed} and shear V_{Ed}. In the erection phase, the steel profiles encased in the columns or walls can serve as temporary supports to formwork; they do not need to be dismantled after pouring concrete, which reduces the construction time.

4.1.3 A specific design issue: longitudinal shear at steel profiles-concrete interfaces

In reinforced concrete, bars must be anchored to transfer forces. The design checks for the anchorage of bars near supports and for overlaps between bars are explicit, but no design check is required for the longitudinal shear at the bar-concrete interface, though this longitudinal shear is also present outside of the support zone in reinforced concrete components submitted to bending and transverse shear. This allowance is justified by the fact that all bars with indentations offer a resistance to longitudinal shear that always surpasses the demand.

There are no indentations in the surface of steel profiles, and their bond resistance per unit surface is small, of the order of seven times lower than the bond strength of ribbed bars with a C30 concrete and more than seven times for higher concrete classes. Although profiles exhibit a larger surface to develop shear resistance by bond, this does not compensate for their low bond strength. This can be illustrated in the following example: applying Eurocode 2 (2004), the design value of the ultimate bond stress f_{bd} for ribbed bars in a C30 concrete, if the bond condition and the position of the bar during concreting are bad, may be estimated at 2.1 N/mm². For an encased steel H section, the resistance to longitudinal shear τ_{Rd} given by Eurocode 4 (2004) is 0.30N/mm², which represents only 14% of f_{bd}. The parameter which defines the ability of a steel section and of a bar to transfer forces by bond are respectively:

$$R_{profile} = [\text{section perimeter} \times \tau_{Rd}] / (\text{section area} \times f_y) \tag{4.1a}$$

$$R_{bar} = [\text{section perimeter} \times f_{bd}] / (\text{section area} \times f_y) (\tag{4.1b}$$

For an S500 bar of diameter 20 mm, R_{bar} = 0.00084; for a HE300B profile made of S460 steel, $R_{profile}$ = 0.000085, approximately ten times less. The ratio R is thus much less favourable for steel sections than for bars and shear connectors may be necessary. It is not possible with encased steel profiles to escape an explicit control of resistance to longitudinal shear at the steel profile-concrete interface in the current length of hybrid walls or columns. Accordingly, controlling the resistance of steel profile anchorages is of paramount importance where this is required, typically in the zones of force application or at supports.

4.2 LOAD-DEFLECTION BEHAVIOUR OF REINFORCED CONCRETE WALLS

Hybrid walls are a particular case of reinforced concrete walls. Therefore it is useful to recall the characteristics of the load-deflection behaviour of reinforced concrete walls. They will serve as a basis for hybrid walls.

4.2.1 Characterization of load-deflection diagrams of reinforced concrete walls

The deformation capacity in rotation of walls that yield in flexure, with or without axial force, may be characterized by the chord rotation or, equivalent, the component drift ratio θ defined as the deflection Δ at the end of the shear span divided by the shear span L_V, where L_V is equal to the ratio M/V of the moment M to the shear V at the point of contraflexure (the base section in cantilever walls) – see Figure 4.4. The reference line considered to define the drift Δ is the tangent to the axis of the component at the yielding section. In cantilever walls, the shear span L_V is the distance H_w between the point of application of V and the base section and θ = Δ/H_w. This way to characterize the deformation capacity in rotation of

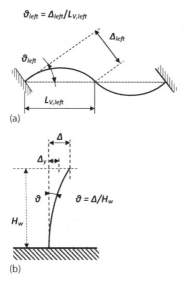

Figure 4.4 a, b Definition of the chord rotation θ: (a) in a frame component; (b) in a cantilever

walls that yield in flexure is common to FEMA356, ACI318-19, AISC341-16 and Eurocode 8 (2022).

Other parameters are currently used in the assessment of wall deformation: Φ_y yield curvature of the base section as established by a section analysis; M_y yield moment of the base section; Q_y shear force which generates M_y; θ_y chord rotation of the shear span at yielding of the end section; k_{eff} wall effective stiffness; θ_u ultimate chord rotation; μ ductility coefficient: $\mu = \theta_u / \theta_y$.

4.2.2 Components of a reinforced concrete wall deflection

Figure 4.5 presents a typical force F, displacement Δ curve of a wall submitted to a combination of bending moment, shear and axial force. In all generality, the displacement Δ at any level in the wall is constituted by a sum of components:

$$\Delta = \Delta_{bending} + \Delta_{shear} + \Delta_{slip} + \Delta_{slide} + \Delta_{anchor} \tag{4.2}$$

$\Delta_{bending}$ is due to flexibility in bending and is always present.

Δ_{shear} is due to flexibility in shear and is always present.

Δ_{slip} is due to the elongation of bars or encased profiles along their anchorage length.

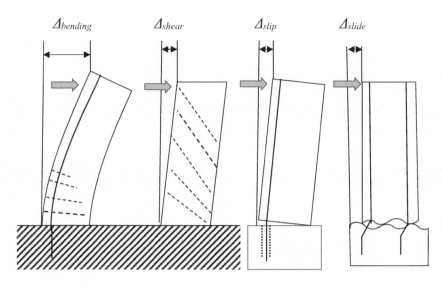

Figure 4.5 Four components of the deflection Δ of a wall. From left to right: flexure, shear, slip and sliding

Δ_{slide} is due to sliding shear along weak planes in the concrete.
Δ_{anchor} is due to the flexibility of the anchorage beam or slab.

4.2.2.1 Flexural deformation

The flexural deformation of relatively slender walls can be assessed by means of the theory which takes into account concrete cracking in tension and steel yielding. This calculation is typically done with a fibre model. For cantilever walls, the diagram of bending moments is statically determined, and the curvature Φ in each section along the component may be found by the analysis of those sections so that a diagram M-Φ can be readily established and a force-displacement diagram M-Δ deduced.

The M-Δ line is a curve with several stages. For low applied moments M, this corresponds to a complete component without cracking. With increased M, cracking appears where tension stresses overpass the resistance of concrete to tension. With further increased M, cracking expands and the area of concrete in compression decreases which induces a progressive reduction of the inertia provided by concrete; the behaviour may still be elastic, but it is non-linear. With further increased M, the inertia keeps getting smaller, a trend which is increased by the progressive yielding of the steel and concrete components of the section. Figure 4.24, which shows stages of progression of cracking considered in the simplified determination of diagrams of interaction between bending moment and axial force, shows the evolution of

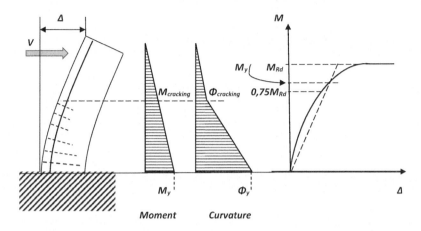

Figure 4.6 Distribution of bending moment and curvature along the height of a wall. Force-displacement diagram of that wall, exact and simplified into a bi-linear representation

cracking described above. Although the exact M-Δ diagram may be available in all its complexity, it is generally preferred, in view of the analysis of structures, to idealize it by means of a bi-linear approximation: an elastic part up to first yield and a plastic part – see Figure 4.6.

The most common way to derive a yield curvature Φ_y from the exact M-Φ diagram in a section is, as shown in Figure 4.7, following details in the literature (Priestley and Kowalsky, 1998; Moehle, 2015). Two straight lines are drawn: one from the origin and which passes at the first yield point $(M_{y1} - \Phi_y')$, and another one which is horizontal at the level of the nominal flexural strength M_{Rd}. The yield curvature Φ_y is found at the intersection A of the two lines. The first yield moment M_{y1} is defined as the moment of first yield of the longitudinal reinforcement or the moment at which the

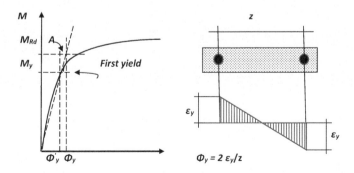

Figure 4.7 Definition of Φ_y' and Φ_y. Relationship $\varepsilon_y - \Phi_y$ for reinforcement concentrated in boundary zones

compressive strain in concrete becomes equal to $\varepsilon_{c2} = 0.002$. The section analysis for M_{y1} gives the curvature Φ_y'.

In the frame of Eurocode 2 (2022), the nominal flexural strength M_{Rd} is found for a compressive strain limit $\varepsilon_{cu} = 0.0035$ in concrete, and a tensile strain limit for reinforcing steel $\varepsilon_{ud}=f_{yd}/E_s$ (if the post-elastic branch for steel is not supposed horizontal, a strain limit $\varepsilon_{ud}\leq0.9\varepsilon_{uk}$ and a maximum stress $k\,f_{yk}/\gamma_s$ at ε_{uk}, where $k = (f_t/f_y)$, are imposed).

The definition of the yield curvature Φ_y can be more direct than described above. In pure bending, the yield curvature Φ_y of a wall is related to the yield stress of the reinforcement placed for bending resistance (Priestley and Kowalsky, 1998). If those reinforcements are concentrated in the boundary zones of a rectangular wall section (see Figure 4.7), then

$$\Phi_y = 2\varepsilon_y/z \tag{4.3a}$$

where z is the distance between the centroids of the boundary zones, and ε_y is, for standard S500 reinforcement, $\varepsilon_y = f_y/E_s = 500/210000 = 0.0024$

The yield strain ε_y of reinforcement being smaller than the limit strain of concrete $\varepsilon_{cu} = 0.0035$, the behaviour of the section can be plastic for strains greater than 0.0024, up to an ultimate curvature of Φ_u which fully exploits the shortening capacity of concrete up to $\varepsilon_{cu}=0.0035$.

If the longitudinal reinforcement is uniformly distributed in a rectangular concrete section (Priestley and Kowalsky, 1998), the yield curvature Φ_y is given by:

$$\Phi_y = 2.4\varepsilon_y/z \tag{4.3b}$$

The factor which multiplies ε_y/z in (4.3a) and (4.3b) is similar to the "shape factor" W_{pl}/W_{el} of steel sections, where W_{pl} is the plastic section modulus and W_{el} is the elastic section modulus; W_{el} corresponds to first yield and W_{pl} to the plastic moment of resistance. The steel equivalent of a reinforced concrete section with distributed reinforcement is a steel rectangle with $W_{pl}/W_{el}=1.4$. The steel equivalent of a concrete section with reinforcement concentrated in the boundary zones is a H section with $W_{pl}/W_{el}=1.14$. The ratio of the shape factors is the same for steel section and concrete walls: $1.4/1.14\approx2.4/2=1.2$. In practice, each of Equations (4.3a) or (4.3b) may be closer to reality, depending on the more or less concentrated density of longitudinal reinforcement, bars or encased profiles.

It can also be noticed that the ratio M_{Rd}/M_{y1}, like the shape factor, is close to 1.0 if longitudinal reinforcement is concentrated in boundary zones; then, Point A of intersection between the two straight lines of the bi-linear approximation of the M-Φ diagram may be defined without significant error as (Φ_y, M_{Rd}).

If an axial force N_{Ed} is applied together with a bending moment M_{Ed}, the axial strain due to N_{Ed} uses part of the available strain capacity of steel and concrete in compression. The yield curvature Φ_y remains unchanged, but the ultimate curvature Φ_u is reduced in comparison to the pure bending case, which results in a reduction of the ductility $\mu = \Phi_u/\Phi_y$ in comparison to the pure bending case. For seismic design, the limitation of the axial force ratio N_{Ed}/Af_c in Eurocode 8 (2022) or the limitation of strains in concrete under axial force in ACI318-19 and AISC341-16 are justified by the wish to keep high enough the available ductility μ.

Once Point A of the bi-linear approximation of the M-Φ diagram is defined, the effective flexural stiffness $k_{eff,M}$ or $EI_{eff,M}$ coherent with bending deformations only is given by:

$$k_{eff,M} = M_{Rd}/\Phi_y = EI_{eff,M} \tag{4.4a}$$

If $k_{eff,M} = EI_{eff,M}$ is assumed constant over the height H_w of a cantilever wall, the deflection $\Delta_{y,bending}$ of the wall at the point of application of the horizontal force Q_y, which corresponds to the moment of resistance M_{Rd} and is only due to bending, may be calculated:

$$k_{eff,M} = Q_y\, H^3/3\Delta_{y,bending} \tag{4.4b}$$

$$\Delta_{bending} = Q_y\, H^3/3EI_{eff,M} = M_yH^2/3EI_{eff} = \Phi_y\, H^2/3 \tag{4.5}$$

The chord rotation θ_y is: $\theta_{y,bending} = \Delta_{bending}/H_w = \Phi_yH/3$ \qquad (4.6)

Some comments can be made on the whole process of characterizing the flexural behaviour of walls and columns.

There is always some approximation in the "exact" calculation of the M-Φ relationship, because the initiation of cracking and the opening of flexural cracks are influenced by other parameters than the action effects M_{Ed} and N_{Ed}, such as concrete shrinkage, the diameter of the reinforcement, creep under gravity loading and the interaction of flexural cracks with shear cracks.

There are many possible definitions of the M-Φ bi-linear approximation. For the elastic part, two have been given above (passing through M_{y1}, estimations of Φ_y). For the plastic part, the level of the plateau may be defined in different ways, and strain hardening may be considered or not. The definition of the M-Φ bi-linear approximation is thus to a certain extent conventional.

The definition of force-displacement Q-Δ bi-linear approximations is also conventional and there are different definitions of Point A, of Q_y and of Δ_y in design codes and the literature. The definitions are related to the perspective in which parameters like stiffness, yield strength and post-yield strain hardening are taken into account: for static design, the objective is to define at best a stiffness representative of the elastic part of the behaviour; in the seismic design context, the objective may be a bi-linear approximation in which the absorbed energy up to a performance point is the same as with the exact force-displacement diagram obtained in cyclic tests; Eurocode 8 (2022) gives such a definition. It is important to stick to a given definition to make valid comparisons between the results of different testing programs.

As a result of the non-linear character of the M-Φ relationship, the elastic part of a Q-Δ bi-linear approximation is necessarily a secant of the exact force-displacement relationship; a straight line M-Φ or Q-Δ tangent at the origin to a calculated exact curve or to an experimental curve should not be used to define a sensible bi-linear approximation.

In the case that an experimental force-displacement Q-Δ relationship is available, the bi-linear Q-Δ relationship is often defined by taking, for the elastic part of the behaviour of reinforced concrete components, the secant passing through the point at which $Q = 0.75 \, Q_{max}$ and in adopting:

$$Q_y = 0.75 \, Q_{max} \tag{4.7}$$

where Q_{max} is the maximum resistance observed in the test. Δ_y and M_y are found accordingly and k_{eff} is calculated with Equation (4.4b).

The plastic part of the bi-linear Q-Δ relationship may be horizontal at the Q_y level up to the displacement at which a drop in resistance of 20% with respect to Q_{max} is observed. That approach generally gives values of Q_y similar to those found in the way explained above, but it must be recalled that the effective stiffness k_{eff} found from the experimental Q-Δ relationship includes the effect of shear and slip deformations so that k_{eff} is smaller than $k_{eff,M}$.

There is also an approximation in Equations (4.3) and (4.4b) in that they ignore the tension shift of the bending moment diagram at the base of the wall. This tension shift is due to the spread of inelastic effects calculated at the base of the wall over a plastic hinge length: tension in reinforcement at a height of the order of the wall length is the same as at the wall base. This effect is taken into account in the complete Eurocode 8 (2022) Equation (4.10) for the calculation of chord rotation θ_y at yield.

4.2.2.2 Shear deformation

Shear deformations can be estimated by idealizing walls as made of a homogeneous isotropic material with constant shear modulus. For a cantilever wall submitted to Q_y the top displacement due to shear may be found as:

$$\Delta_{shear} = Q_y \, H_w \, / \, A_v G_{eff} \tag{4.8}$$

where A_v is the shear area of the section, and G_{eff} is the effective shear modulus of concrete.

Before shear cracking, the shear modulus G_c of concrete is $G_c=0.4E_c$. As explained in Elwood and Eberhard (2006), the modulus decreases with increasing shear down to:

$$G_{eff}=0.2E_c \qquad (4.9)$$

Concrete in service is not a homogeneous isotropic material, and a model closer to reality is developed in Section 4.5 for hybrid walls and columns. Δ_{shear} may be much greater than $\Delta_{bending}$ in walls of low slenderness, which means walls with an aspect ratio H/l_w below 1.5.

4.2.2.3 Slip deformation

At the base of a wall, the stress in the flexural reinforcement is high, up to f_y. The tension force is progressively transferred into the concrete through bond stress and decreases down to zero at the end of the anchorage length, where the bars may be considered fixed. The elongation of the bars between that point and the top of the anchorage block results in a rotation of the wall considered as a rigid body for this component Δ_{slip} of the deflection. The elongation is related to the properties of the steel bars and of the surrounding concrete as indicated by the shear contribution to θ_y in Equation (4.10).

4.2.2.4 Deformation of the anchorage block

The flexibility of the anchorage beam or slab of a wall which induces a deflection Δ_{anchor} is normally taken into account in the analysis of a structure.

In tests, Δ_{anchor} should always be measured and, to avoid its interference in the terms characterizing the intrinsic flexibility of the wall, Δ_{anchor} should be deduced from the total displacement Δ measured at the load application level before any other data processing made in order to calculate wall stiffness. If this is not done, the stiffness of the wall will be underestimated.

4.2.2.5 Sliding shear deformation

Sliding due to shear along weak planes in the concrete, typically at interfaces between concrete poured at different times, in particular at the base of a wall, can induce a global displacement Δ_{slide} of a wall. Sliding is evitable by a surface preparation of the firstly poured concrete and by dowels crossing the weak planes. This may be assessed by referring to code rules. Encased steel profiles largely contribute to preventing sliding shear.

4.2.3 Calculation of the total deformation of RC walls up to flexural yielding

There are two possibilities to calculate the total deformation up to flexural yielding of a wall that yields in flexure for a bending moment M_y generated by an applied force Q_y:

- Either by calculations making use of an equation that explicitly gives the flexure, slip and shear components of a wall deformation.
- Or by calculations making use of an effective stiffness k_{eff} calibrated to take into account those three components of deformation.

4.2.3.1 Calculation of flexure, slip and shear components of wall deformation at yield in flexure

Equations have been established that explicitly give the flexure, slip and shear components of the deformation. In Eurocode 8 (2022), the chord rotation θ_y at yielding of rectangular walls may be evaluated using Equation (4.10); the term for flexure applies to Equation (4.6), taking into account both the shear span L_v and the tension shift a_1 of the bending moment diagram:

$$\theta_y = \phi_y \frac{L_V + a_1}{3} + \frac{\phi_y d_{bL} f_y}{8\sqrt{f_c}} + 0.0011\left(1 + \frac{l_w}{3L_V}\right) \text{ (in rad)} \qquad (4.10)$$

where l_w is the section depth, f_y the longitudinal steel mean yield strength in MPa, f_c the concrete mean compressive strength in MPa and d_{bL} is the mean diameter of the tension reinforcement; the notations Φ_y and L_V are defined in 4.2.1; all lengths are in mm.

If the transverse reinforcement is horizontal, the tension shift a_1 of the bending moment diagram may be found as $a_1 = z \cot\theta/2$ where θ is the inclination of the concrete compression struts over the wall vertical axis and z is the internal lever arm of forces; z is taken equal to $d + d'$ in walls with barbelled or T-section, with d and d' the depths to the tension and compression reinforcement, or $z = 0.8\ l_w$ in walls with rectangular sections.

4.2.3.2 Calculation of wall deformation by means of an effective stiffness

The second possibility to calculate the total deformation of a wall up to flexural yielding uses expressions of the effective stiffness k_{eff} calibrated on tests and which generally include the effects of flexure, slip and shear, acting together, though some references like FEMA356 separate the effect of shear. Effective stiffness also takes shrinkage into account, in an approximate way, because cracks due to shrinkage are always present and reduce the stiffness effectively available before any application of a bending moment or an axial force; this for instance explains the factor 0.8 applied to $E_c I_g$ in Equation (3.2).

Effective stiffness k_{eff} may be expressed either in the form $\alpha E_c I_g$, where E_c is the Young's Modulus of concrete and I_g the inertia of the gross concrete section of the wall, or in the form of a combination of terms related to, respectively, the stiffness of concrete sections, of reinforcement by bars and of reinforcement by encased profiles – see Equation (3.2). For both types of equations, variability around the proposed values is inevitable because the parameters characterizing the physics of the problem are not all present in the equation.

Estimates k_{eff} of the stiffness of components based on the external dimensions of reinforced concrete sections are useful for a first calculation of the design action effects M_{Ed}, N_{Ed} and V_{Ed} in a structure because this can be done before knowing the steel reinforcement.

Consideration of the factors influencing the stiffness may be different for design checks related to strength, stability or resistance to earthquakes: considering creep is required for long-duration loading; considering the reduction of stiffness due to crack openings is vital for stability checks – see Chapter 3; it is also necessary for seismic design because the stiffness of components influence the periods of vibration of structures and, as a consequence, the design action effects.

For estimates of stiffness based on the external dimensions of concrete, design codes propose a different coefficient α_{EI} to define effective stiffness in the form $k_{eff} = E_c I_{eff} = \alpha_{EcIg} E_c I_g$. All the design equations hereunder are intended for design under design loads (factored loads).

For concrete walls reinforced by bars, it is indicated in FEMA356 (2000) that k_{eff} may be taken as $0.8\,E_c I_g$ in uncracked walls and $0.5\,E_c I_g$ in cracked walls, but deformations due to shear should also be calculated, with a shear stiffness equal to $0.4\,E_c A_w$. For concrete columns, FEMA356 relates k_{eff} to axial compression levels: k_{eff} may be taken as $0.7\,E_c I_g$ if compression due to design gravity loads is greater than $0.5\,A_g f_{cd}$ and as $0.5\,E_c I_g$ for smaller gravity load or if the component is in tension. Comparison of experimental results with indications in FEMA356 (2000) – see Figure 4.8 – have shown that k_{eff} may be significantly smaller than $0.5\,E_c I_g$, and this is taken into account in more recent documents.

For reinforced concrete walls, ACI 318-19 gives, for an elastic analysis of a structure, the choice between simple and more complete equations and imposes lower and upper bounds to k_{eff}. The simple equation for the in-plane stiffness of uncracked walls under design loads is $k_{eff}=0.7 E_c I_g$ and for cracked walls, $k_{eff} = 0.35 E_c I_g$. The complete equation is:

$$k_{eff} = [0.80+25(A_s)/A_g][1 - M_{pl,Rd}/(N_{pl,Rd} \times l_w) - 0.5 N_{Ed}/N_{pl,Rd}] \times E_c I_g$$

The lower and upper bounds are $k_{eff} \geq 0.35\,E_c I_g$ and $k_{eff} \leq 0.875\,E_c I_g$.

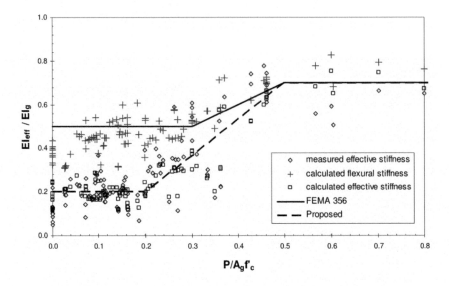

Figure 4.8 Test results and code definitions of the effective stiffness k_{eff} in Elwood et al. (2006)

For seismic design, ACI 318-19 indicates, for components near the onset of yielding, $k_{eff}=0.35E_cI_g$, to be considered together with a shear stiffness $0.2A_gE_c$.

For composite shear walls, AISC341-16 conforms to ACI 318 simple and more complete equation and to the upper and lower bounds for k_{eff}. The more complete equation is written as

$$k_{eff} = [0.80+25(A_a+A_s)/A_g][1-M_{pl,Rd}/(N_{pl,Rd} \times l_w)-0.5N_{Ed}/N_{pl,Rd}] \times E_cI_g$$

Eurocode 8 (2004) proposes a single value: $k_{eff} = 0.5 E_cI_g$, but other factors than 0.5 may be used, if justified, and the more recent Eurocode 2 (2022) indicates $k_{eff} = 0.4 E_cI_g$ for second-order elastic analysis.

4.2.4 Calculation of the ultimate chord rotation θ_u of reinforced concrete walls

Eurocode 8 (2022) gives a way to calculate the ultimate chord rotation θ_u of a critical zone at the end of a concrete component. θ_u is found as the sum $\theta_y + \theta_u^{pl}$ where θ_u^{pl} is the plastic part of the chord rotation, found as the product of a reference θ_{u0}^{pl}, equal to 0.023 rad for rectangular walls or 0.027 rad for other walls, by a set of factors taking into account concrete strength other than 25 MPa, shear span other than 2.5, type of detailing for the boundary zones and axial force other than 0.

4.3 CALCULATION OF SHEAR ACTION EFFECTS IN HYBRID WALLS AND COLUMNS

4.3.1 Introduction

A straightforward analytical method to calculate the transverse shear in encased profiles and the longitudinal shear between profiles and concrete in a hybrid wall or column submitted to an axial force N_{Ed}, a bending moment M_{Ed} and a shear force V_{Ed} is needed for daily practice in order to check the encased profiles and to design the shear connectors which may be necessary at the steel profiles-concrete interfaces. Two practical analytical methods are available; one is an application of the classical Euler-Bernoulli elastic beam theory, and the other one is based on an extension of the truss model of reinforced concrete. The choice has been made to present both of them for two reasons:

- Due to the difference in the hypothesis on which they are based, each method applies best in different ranges of N_{Ed} and M_{Ed} combined action effects.
- Both methods are used in practice.

4.3.2 Classical beam model

The action effects in a hybrid wall or column submitted to an axial force N_{Ed} and a bending moment M_{Ed} may be determined by means of the classical elastic beam theory (Plumier et al., 2012). That approach which is explained in Section 4.4 has one drawback: it is based on the hypothesis of elastic behaviour of components that are made of materials resisting tension and compression, without cracking; accordingly, principal compression and tension stresses induced by a transverse shear force acting alone are inclined at 45° of the axis of the wall or column. These hypotheses do not generally correspond to concrete components, but they may be close enough to reality in walls or columns submitted essentially to compression and marginally to bending and shear because, in that case, tensile bending stresses are smaller than compression stresses, which they only reduce elastically, and cracking affects only a very limited part of the section. It results that the beam model covers best the design for longitudinal shear of walls or columns submitted essentially to compression and marginally to bending and transverse shear. Detailed examples of its application to different types of sections may be found in the literature, e.g. in Bogdan et al. (2019).

4.3.3 Elastic models taking into account shear deformation of sections

The Timoshenko model of the complete elastic theory of beams (Timoshenko, 1951) takes into account shear deformation of beams by which plane sections do not remain plane. It is not always necessary to take into account this effect:

it has been shown experimentally, for instance in Zhou et al. (2010) or Zhang et al. (2020), that plane sections remain plane in tests if structural conditions which exist at the storey level in real buildings are correctly reproduced in the test set-up, which means if the storey forces are introduced in the wall by means of a reinforced concrete beam part of the wall and if the reaction at the base of the wall is provided by a stiff reinforced concrete foundation block.

However, for beams with encased steel profiles in which both the load application and the reaction force are applied punctually on the short side of the component, it may be more realistic to take into account the non-plane deformations of initial plane sections. This is also necessary if the shear connection at steel profiles to concrete interfaces are semi-rigid partial strength connections. A model of reinforced concrete components such as Vierendeel beams developed in Chrzanowski (2019) takes into account such situations. Being not statically determinate, the model is more complex to use than the classical beam theory or the hybrid truss model; its calibration, at present realized on only two specimens submitted to pure bending, requires complementary work.

4.3.4 Hybrid truss model

A model alternative to the application of the classical or complete elastic beam theory is proposed in this chapter. It extrapolates the Ritter-Mörsch truss model to the case of walls with embedded steel profiles, a fundamental ultimate state model for reinforced concrete design which is internationally recognized and serves as a reference for design expressions in most design codes: Eurocode 2 (2004); ACI318-19; Eurocode 2 (2022); etc.). For detailed explanations of the Ritter-Mörsch truss model, see for instance Grandić et al. (2015). The model proposed in this chapter, referred to as the "hybrid truss model", is explained in Section 4.5. It fits in the frame of the usual reinforced concrete design practice and, as shown in Section 4.9, gives results which compare successfully with experimental observations. It was first proposed in Plumier, A. (2015) and in Plumier et al. (2016). The method applies without the restrictions related to the elastic beam model, in particular to columns or walls submitted to significant bending and shear effects and to relatively low compression forces: in that case, cracking affects a significant part of the section, which corresponds to the hypothesis of the hybrid truss model.

4.4 CALCULATION OF THE SHEAR ACTION EFFECTS IN HYBRID COMPONENTS BY THE ELASTIC BEAM MODEL

4.4.1 Longitudinal shear at concrete-steel profiles interface

In the theory of elastic beams, the longitudinal shear force $V_{Ed,l}$ per unit longitudinal length of wall in one plane cut like bb at Figure 4.9, is equal to:

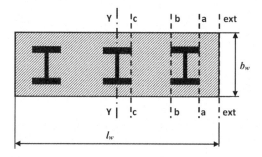

Figure 4.9 Section plane cuts to calculate the longitudinal shear $V_{Ed,l}$

$$V_{Ed,l} = V_{Ed}\, S/I \qquad\qquad\qquad (4.11)$$

where V_{Ed} is the applied transverse shear, S is the first moment of the area comprised between the exterior side *ext-ext* of the section and the plane cut bb, with reference to the neutral axis YY of the section. I is the second moment of area (inertia) of the complete section about YY.

To calculate the longitudinal shear force $V_{Ed,l}$ at the interface concrete-steel profile in the application of the elastic beam model, the more direct way consists in applying Equation (4.11) to a part of the complete section which has the width of an encased profile. To do so, the wall or column sections should be divided into "subsections", some being hybrid and having the width of encased profiles, and the other ones being reinforced concrete (Figure 4.10). Then it is necessary to calculate the part $V_{Ed,i}$ of the total shear V_{Ed} which is applied to each subsection. Since all subsections work together in a single solid, they all have the same deformed shape over the column length, which implies that the

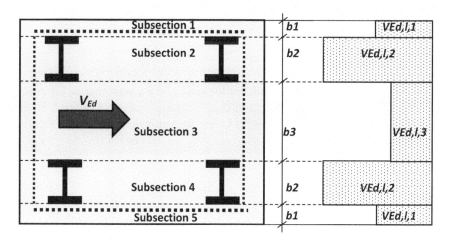

Figure 4.10 Distribution of transverse shear V_{Ed} in a section of hybrid column, with hybrid subsections of width $b2$ and reinforced concrete subsections of width $b1$ and $b3$

action effects allocated to each subsection are proportional to the parameter that most affects its deformed shape: its effective flexural stiffness EI_{eff}. In fact, the shear stiffness also intervenes but in a secondary manner, and, considering EI_{eff}, is accurate enough for this subdivision operation. The transverse shear $V_{Ed,i}$ applied to each subsection i is thus proportional to the relative flexural stiffness i of each subsection:

$$V_{Ed,i} = V_{Ed} \times EI_{eff,i}/\Sigma EI_{eff,i} \tag{4.12}$$

In each subsection, the components of $EI_{eff,i}$ due to rebars, to concrete and to steel profiles may be combined, using for instance Equation (3.2) of effective flexural stiffness $(EI)_{eff,II}$ of Eurocode 4 (2004) reproduced hereunder:

$$EI_{eff} = (EI)_{eff,II} = K_0 (E_a I_a + E_s I_s + K_{e,II} E_{cm} I_c) = 0.9 (E_a I_a + E_s I_s + 0.5 E_{cm} I_c)$$

Note: for simplicity, notation EI_{eff} is used hereunder in place of $(EI)_{eff,II}$.

The application of Equations (3.2) and (4.12) provides a distribution of bending stiffness and of transverse shear of the type shown in Figure 4.10, which shows the major contribution of the hybrid subsections to the global flexural stiffness EI_{eff} and the major transverse shear in those subsections. Design checks of shear resistance have to be made separately in the different subsections because these are submitted to different levels of shear. The design resistance to shear should be, in each subsection, greater than the action effect $V_{Ed,i}$ in that subsection. The stirrups that are necessary in one subsection should be installed in or just around that subsection.

To calculate with Equation (4.11) the longitudinal shear in cuts like aa of Figure 4.9, each composite subsection should be "homogenized" in only steel or only concrete. If subsections are homogenized into concrete, the moment of inertia I_c^* of a hybrid subsection like bs should be such that $E_{cm} I_c^*$ is equal to the stiffness $(EI_{eff})_{bs}$ of that subsection:

$$I_c^* = (EI_{eff})_{bs}/E_{cm} \tag{4.13}$$

Considering the effective modulus for steel and for concrete in Equation (3.2), which are respectively E_s and $0.5E_{cm}$, compliance to Equation (4.13) requires that the width t_f^* and h_w^* of concrete equivalent respectively to the width t_f of steel flanges and to the height h_w of the steel web in a concrete homogenized section be calculated as:

$$t_f^* = t_f \times E_s/(0.5E_{cm}) \quad h_w^* = h_w \times E_s/(0.5E_{cm}) \tag{4.14}$$

where E_{cm} is the secant modulus for short-term loading.

The resultant longitudinal shear force per unit length in sections like C1 and C2 in Figure 4.11 is calculated with Equation (4.11) written as $V_{Ed,l} = (V_{Ed,bs} \times S) / I_c^*$ in which S is the first moment of the areas between sections

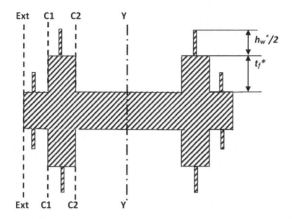

Figure 4.11 Homogenized concrete section equivalent to the hybrid subsection of width
b2 of Figure 4.10

C1 or C2 and Ext, the upper face of the section in Figure 4.6. Resistance to longitudinal shear should be checked on the concrete width at interfaces C1 and C2 between steel profiles and concrete.

The approach described hereabove is purely elastic. Keeping the hypothesis of a beam behaviour in which plane sections remain plane, but envisaging a plastic limit state, allows a simpler calculation of the longitudinal shear if the composite section is mainly submitted to bending. Based on the definition of the longitudinal shear, it can be concluded that the maximum possible shear forces developed in plane sections like CC1 or CC2 of Figure 4.11 are equal to the tensile/compression resistance of the weaker composing part above or below the considered plane section. Figure 4.11 shows the plastic strength of the concrete section above CC1 for the design check at CC1 and the plastic strength of the concrete plus steel sections for the design check at CC2. It is a common practice for the evaluation of the required shear connection resistance in beams to be set as equal to the aforementioned tensile/compression resistance. Like this, the number of required shear connectors can be easily evaluated in a conservative way; this possibility is mentioned in Eurocode 4 (2004).

4.4.2 Transverse shear in steel profiles

Transverse shear in steel profiles influences their available strength in tension or compression and this influences the capacity of the hybrid section in bending. This has to be taken into account in a way explained in Section 4.6.4.

In the elastic beam model, it may be considered that the transverse shear $V_{Ed,i}$ in a subsection is distributed among the components of the subsection proportionally to their shear stiffness:

The shear stiffness K_{SP} of the set of encased steel profiles is equal to NG_sA_v, where N is the number of profiles in the hybrid subsection, G is the shear modulus of steel (80,769 MPa) and $A_{v\,is}$ the shear area of one profile. The shear stiffness K_{RC} of the concrete is equal to G_cA_c, where A_c is the area of concrete in the hybrid subsection and G_c is the shear modulus of concrete; G_c is related to E_c by the expression $G_c = E_{cm}/[2 \times (1 + \nu_c)]$ where ν_c is the Poisson's coefficient of concrete. However, like always with concrete, the value of E_{cm} or E_c depends on cracking. In sections essentially submitted to compression, G_c may be taken as $G_c = 0.4E_{cm}$. In sections essentially submitted to bending, G_c may be taken as $G_c = 0.2E_{cm}$.

The design transverse shear $V_{Ed,a}$ acting in one steel profile, necessary to assess the possible reduction of strength of that profile in the longitudinal direction as explained in Section 4.6.4, is:

$$V_{Ed,a} = V_{Ed,i} \times K_{SP} / [N(K_{SP} + K_{RC})] \tag{4.15}$$

4.5 CALCULATION OF THE SHEAR ACTION EFFECTS IN HYBRID COMPONENTS BY THE HYBRID TRUSS MODEL

4.5.1 The hybrid truss model concept for the analysis of hybrid walls

As explained in Section 4.2, there are five terms contributing to the total deflection of hybrid walls subjected to shear and bending. If we focus on the terms due to the internal deformations of the wall itself, the total deflection $\Delta_{tot,MV}$ is the sum of two components: the bending component $\Delta_{bending}$ and the component due to shear Δ_{shear}, as illustrated in Figure 4.12:

$$\Delta_{tot,MV} = \Delta_{bending} + \Delta_{shear} \tag{4.16}$$

In the truss analogy applied to reinforced concrete, all bars contribute to the truss stiffness by their axial stiffness EA and the individual bending stiffness EI of the bars is considered negligible in comparison to the bending stiffness of the truss provided by the axial stiffness of the chords of the truss and by the lever arm between the tension and compression chords. It can be shown that this also generally applies to hybrid sections. For instance, in the wall section of Figure 4.23, the stiffness EI of the three encased HE120B sections represent only 2% of the wall effective stiffness EI_{eff}. This 2% difference is negligible and may be ignored, so that the methods of analysis used for reinforced concrete components in bending may be used with hybrid components, in the elastic stage as well as in the plastic stage. The practical implication is that steel profiles can be assimilated to bars for the calculation

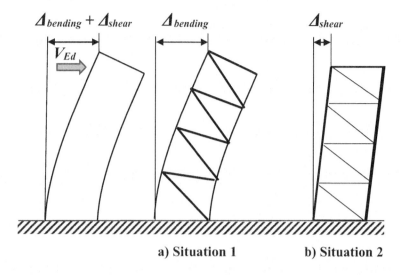

a) Situation 1   b) Situation 2

Figure 4.12 Components of the deformation of walls. Thick lines indicate components with infinite stiffness

of resistance to bending moments M_{Ed} and axial forces N_{Ed}. In the truss analogy applied to reinforced concrete, the individual shear stiffness GA of the bars is also considered negligible in comparison to the shear stiffness of the truss provided by the axial stiffness of the concrete compression struts and of the steel ties. In reinforced concrete, the diagonals and ties of the truss are designed such that the truss provides resistance to transverse shear. But, in hybrid components, the shear stiffness and strength brought by encased steel profiles are not negligible: if the shear stiffness and strength due to reinforced concrete in a hybrid wall were zero, the shear stiffness and resistance of the encased profiles would be present anyway.

The concept of a hybrid truss explained in the following takes into account the contribution of the encased profiles to shear stiffness and to resistance to transverse shear. It is based on the following observations.

In reinforced concrete, the truss model is the same for bending action effects and for shear action effects, but these effects are treated separately in two situations that allow the identification of the respective contributions of shear and bending deformations to the total deformation of the truss. In situation 1 (see Figure 4.12 a), the diagonals and transverse bars are assumed to be axially infinitely stiff and the horizontal displacement at the top of the truss is only due to the axial deformation of the chords: $\Delta = \Delta_{bending}$. It is a "bending flexibility only" situation. In situation 2 in Figure 4.12 b, the chords are axially infinitely stiff and the deformation is only due to the axial deformation of the diagonals and transverse bars: $\Delta = \Delta_{shear}$. It is a "shear flexibility only" situation.

The total of situations 1 and 2 includes the flexibility of each component only once in the truss representing the reinforced concrete wall. The axial forces in the bars of the truss are the same in situations 1 and 2 because the truss system is statically determinate.

In a hybrid wall, the encased steel profiles are longitudinal reinforcement parts of the chords, and they are taken into account as such in the calculation of $\Delta_{bending}$. The shear stiffness NGA_s of the N steel profiles, which is not taken into account by the truss model, must be introduced separately. It can be in the form of shear components working in parallel with the truss – see Figure 4.13. In that model, the encased steel profiles are present in the truss by their axial stiffness and strength; they are present in the shear components by their shear stiffness and strength; the truss and the shear components are supposedly connected to the truss by means of infinitely stiff hinged rods, as shown in Figure 4.13. A total shear V_{Ed} applied to a hybrid wall is distributed in the shear components and in the truss proportionally to their relative shear stiffness. The total of shear components and truss is the hybrid truss model.

If K_{RC} is the shear stiffness of the reinforced concrete truss and K_{SP} the shear stiffness of the set of steel profiles, the total transverse shear force $V_{Ed,SP}$ applied to the N-encased steel profiles is found as:

$$V_{Ed,SP} = V_{Ed} K_{SP}/(K_{SP} + K_{RC}) \qquad (4.17)$$

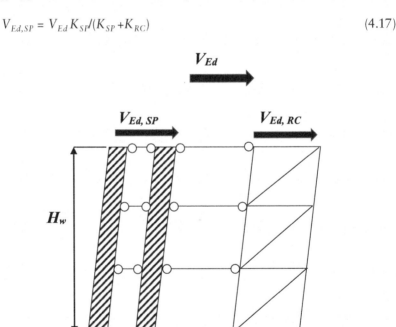

Figure 4.13 Distribution of the total shear V_{Ed} in two shear-resisting systems working in parallel: at left, two encased steel profiles; at right, one reinforced concrete wall figured by a truss

The shear force applied to the reinforced concrete truss is:

$$V_{Ed,RC} = V_{Ed} K_{RC}/(K_{SP} + K_{RC})$$ (4.18)

The way to calculate K_{RC} and K_{SP} is developed in Sections 4.5.2 and 4.5.5. A way to take into account the fact that the steel profiles interact with the concrete compression diagonals of the reinforced concrete truss is proposed in Section 4.5.3.

4.5.2 Shear stiffness K_{RC} of the reinforced concrete truss

The evaluation of the shear stiffness K_{RC} of the reinforced concrete truss is made on a "reference cell" of the truss. This evaluation is classical in reinforced concrete (Moehle, 2015). In hybrid walls, due to the predominant contributions to the bending strength of the encased sections, the distance z between the resultant compression and tension forces in the chords of the hybrid truss is considered as being the distance between the centre of the steel profiles of the chords. Those profiles are referred to as the "external" profiles, and the profiles in the web of the wall are referred to as the "internal" profiles. θ is the angle between the concrete compression diagonals and the chords. The height of the wall corresponding to a "reference cell" is z $cot\theta$ (Figure 4.14).

For an applied shear $V_{Ed,RC}$, the horizontal displacement δ of the application point of $V_{Ed,RC}$ may be found as a function of the characteristics of the truss components by expressing that the energy W_{int} of deformation of the truss components is equal to the work W_{ext} of the applied external force

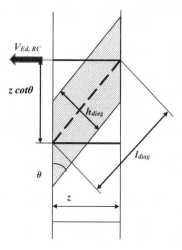

Figure 4.14 "Reference cell" of Mörsch truss subjected to a shear force $V_{Ed,RC}$

$V_{Ed,RC}$; the terms in W_{int} that correspond to the deformation of the chords of the truss are negligible because in "situation 2" envisaged here (see Figure 4.12b), the chords are assumed infinitely stiff:

$$W_{ext} = \frac{V_{Ed,RC}\delta}{2}$$

$$F_{diag} = \frac{V_{Ed,RC}}{sin\theta} \quad A_{diag} = b_w z cos\theta \quad l_{diag} = \frac{z}{sin\theta}$$

$$F_{stirrups} = V_{Ed,RC} A_{stirrups} = A_s = nA_{sw} = \frac{zcot\theta}{s} A_{sw} \quad l_{stirrup} = z$$

$$W_{int} = \sum_i \frac{N_i^2}{2E_iA_i} l_i = \frac{V_{Ed,RC}^2 \, z}{2(sin\theta)^2 \eta E_{cm} b_w z cos\theta sin\theta} + \frac{V_{Ed,RC}^2 z}{2A_s E_s}$$

where F_{diag}, A_{diag}, and l_{diag} are the axial force, the section area and the length of a diagonal, respectively; $F_{stirrups}$, $A_{stirrups}$, and $l_{stirrups}$ are the axial force, the section area and the length of the stirrups present in the "reference cell", respectively; b_w is the wall thickness; $zcos\theta$ is the width of the concrete compression diagonal; s is the step or distance between consecutive stirrups; A_{sw} is the total section area of one stirrup; n and A_s are the number of stirrups and the total section area of all stirrups in the "reference cell", respectively; η is a coefficient, which, as explained in Section 4.5.3, takes into account that encased profiles constitute "hard spots" in the compression diagonal; $\eta=1.0$ in reinforced concrete walls and $\eta > 1.0$ in hybrid walls with one or more encased steel profiles in the web of the wall.

The horizontal displacement δ of the application point of V_c of the reference cell is found by expressing that $W_{int} = W_{ext}$:

$$\delta = V_{Ed,RC}\left(\frac{s}{E_s A_{sw} cot\theta} + \frac{1}{\eta E_{cm} b_w (sin\theta)^3 cos\theta} \right) \quad (4.19)$$

The shear stiffness $K_{RC,ref}$ of the reference cell is found by expressing that $V_{Ed,RC}=1$:

$$K_{RC,ref} = \frac{1}{\delta} = \frac{1}{\dfrac{s}{E_s A_{sw} cot\theta} + \dfrac{1}{\eta E_{cm} b_w (sin\theta)^3 cos\theta}}$$

The shear stiffness K_{RC} per unit vertical length of wall is found by dividing $K_{RC,ref}$ by the vertical length $zcot\theta$ of the reference cell:

$$K_{RC} = \frac{z\eta E_{cm}E_s b_w A_{sw}\left(\sin\theta\right)^2\left(\cos\theta\right)^2}{sb_w\eta E_{cm}\left(\sin\theta\right)^4+E_s A_{sw}}$$

which may be written in a more compact form:

$$K_{RC} = \frac{z\eta E_s b_w \rho_t\left(\sin\theta\right)^2\left(\cos\theta\right)^2}{\eta\left(\sin\theta\right)^4+n_0\rho_t} \tag{4.20}$$

where $n_0 = E_s/E_{cm}$ and ρ_t is the percentage of transverse steel: $\rho_t = A_{sw}/sb_w$

The parameter that most influences K_{RC} is the content ρ_t of transverse steel. If, for instance, $\rho_t = 0.5\%$, $\theta=45°$, $n_0=7$ and $\eta=1$, then $n_0\rho_t = 0.035$ and is negligible in comparison to $(\sin\theta)^4=0.25$; then K_{RC} may be simplified into $K_{RC} = z\,E_s\,b_w\rho_t$.

For other values of θ, K_{RC} may be simplified into $K_{RC} = z\,E_s\,b_w\rho_t\,(tan\,\theta)^2$.

The shear stiffness K_{RC} depends on the inclination θ of the concrete compression struts chosen by the designer. This choice is discussed in Section 4.5.4 and the relevant consequences.

4.5.3 Contribution of encased profiles to the stiffness K_{RC} of compression struts: factor η

The steel profiles cross the concrete compression struts of the truss model. As for the transverse reinforcing bars, no interaction is considered other than taking into account that encased steel profiles are "hard spots" inside the compression struts (see Figure 4.15).

The influence of internal steel profiles on the stiffness of compression struts may be expressed by a coefficient η applied to the reinforced concrete terms in K_{RC}. In the case of internal profiles with a web perpendicular to the wall's strong bending axis (Figure 4.16a), η may be assessed easily, looking at a horizontal section in the wall and doing the following simplifications:

- The flanges of the profiles do not contribute to an increase in the stiffness of the concrete compression struts and all internal profiles have the same section.

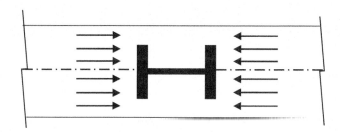

Figure 4.15 Encased steel profiles are "hard spots" in the compression struts

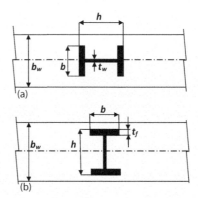

Figure 4.16 a, b Internal steel profile with: at left, a web perpendicular to the wall strong bending axis; at right, a web parallel to the wall strong bending axis

- The webs of the profiles contribute to an increase in the stiffness of the concrete compression struts, and this is expressed by an equivalent concrete width $(E_s/E_{cm})\, t_w$ over the height h of each profile (t_w is the thickness of the web of the profile).
- The stiffness of the concrete section equivalent to the web of the profile over the height h of each profile then is: $E_{cm} h\, [(b_w - t_w) + t_w\, E_s/E_{cm}]$.
- To take into account the contribution to stiffness of the flanges of the profiles and of the confinement of concrete between the flanges, the equation above may be simplified into $h\,(E_{cm} b_w + E_s\, t_w)$.
- The stiffness of the compression strut then is

$$E_{cm} b_w\,(z - N_i\, h) + N_i\, h\,(E_{cm} b_w + E_s\, t_w) = E_{cm} b_w\, z + E_s\, N_i\, h\, t_w$$

where N_i is the number of internal profiles.

- η is the ratio of the stiffness of the compression strut with and without the N_i internal steel profiles:

$$\eta = 1 + \frac{N_i n_0 h t_w}{b_w z} \tag{4.21}$$

In the case of N_i internal profiles with a web parallel to the wall's strong bending axis (Figure 4.16 b), η is similarly found as

$$\eta = 1 + \frac{2 N_i n_0 b t_f}{b_w z} \tag{4.22}$$

For encased concrete-filled steel tubes, η may be taken equal to:

$$\eta = 1 + \frac{N_i n_0 D t}{b_w z} \tag{4.23}$$

where D and t are the diameter and the thickness of the steel tube, respectively.

4.5.4 Inclination θ of the compression struts in the calculation of K_{RC} for design

In the calculation of the design resistance to shear of concrete components, there is flexibility in the choice of the inclination θ of the concrete compression struts.

In Eurocode 2 (2004), a limitation is imposed to θ $(21.8° \leq \theta \leq 45°)$, and the choice of θ influences the required transverse steel content $\rho_t = A_{sw}/sb_w$ because the resistance $V_{Rd,s}$ is a function of those two parameters:

$$V_{Rd,s} = \frac{A_{sw}}{s} zf_{ywd}\cot\theta \qquad (4.24)$$

If θ decreases, $\cot\theta$ increases and, to realize the same required $V_{Rd,s}$, the required transverse steel content A_{sw}/sb_w may be decreased. Whatever the choice of θ, a statically determinate design is defined, but the chosen θ may be different from what a more refined analysis would settle as the real θ because the real field of stresses in the component tends to be the stiffest. In classical reinforced concrete components, the choice of an inclination θ of the compression struts is not critical, because the indentations of bars provide an overabundant resistance to longitudinal shear at the interface between bars and concrete compression struts. With hybrid components, it may be different: the low bond strength provided by a flat steel surface requires a choice of θ as close as possible to reality and which, in any case, leads to a safe-sided design of shear connectors; the applied shear force $V_{Ed,RC}$ allocated to the reinforced concrete should not be underestimated. Equation (4.20) indicates that when $\theta = 30°$, the stiffness K_{RC} is three times greater than for θ equal to $45°$. K_{SP} is independent of θ and Equation (4.18) gives for $\theta = 30°$ a transverse shear action effect $V_{Ed,RC}$ in the reinforced concrete truss which is currently 50% greater than with θ equal to $45°$. Accordingly, the longitudinal shear at the steel-concrete interface is also 50% greater with θ equal to $30°$ than with θ equal to $45°$. The difference would still be greater if θ was chosen smaller than $30°$. For hybrid walls, the choice of a design value for θ is thus a safety issue.

To gain guidance, crack patterns observed in experiments on hybrid walls have been examined. The crack inclination θ has been measured on the pictures and graphs presented in the literature (Dan et al., 2011a,b); Quian et al., 2012; Ji et al., 2015; Cho et al., 2004; Todea et al., 2021; Wu et al., 2018). These documents correspond to a total of more than 50 specimens of cantilever walls anchored in a rigid reinforced concrete basis. The observations can guide the choice of θ for design.

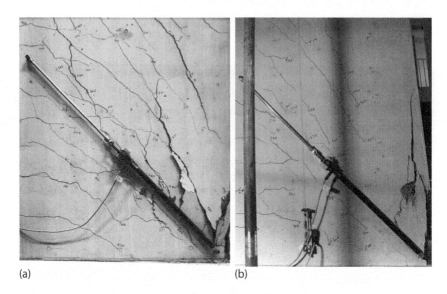

(a)   (b)

Figure 4.17 a, b Crack pattern at failure in two hybrid walls of the SMARTCOCO project at Liege University: at left, specimen BS without shear connectors on the encased profiles; at right, specimen CSN with welded studs on the profiles and high confinement in boundary zones

In walls, for axial load ratios ν_d smaller than 0.32 (in fact in 35 tests out of 50, ν_d is under 0.2), all θ are in a range from 30° to 45° – see for example Figure 4.17. θ equal to 30° is an exception. The most observed value θ is around 35°. In slender walls, θ values around 45° are observed near the base of the wall where the latter is fixed and θ are around 35° above the height of the wall of the order of the wall section height l_w. In the few tests made on squat hybrid walls, θ is on average equal to 45°.

In order to avoid underestimating in design the part of $V_{Ed,RC}$ that is applied to the reinforced concrete truss and on the basis of observations of crack patterns in experiments and of calibration of design equation on experimental results, it is concluded that Equation (4.20) with $\theta=35°$ is adequate for slender walls ($h/l_w \geq 2{,}0$). For squat walls ($h/l_w \leq 1{,}5$), θ should be taken equal to 45°. The practical design equation of the stiffness K_{RC} per unit vertical length of walls then are:

$$K_{RC} = \frac{0.22 z \eta E_s b_w \rho_t}{0.108\eta + n_0\rho_t} \ \left(\text{slender walls, } \theta = 35°\right) \qquad (4.25)$$

$$K_{RC} = \frac{0.25 z \eta E_s b_w \rho_t}{0.25\eta + n_0\rho_t} \ \left(\text{squat walls, } \theta = 45°\right) \qquad (4.26)$$

where $n_0 = E_s/E_{cm}$ and ρ_t is the percentage of transverse steel: $\rho_t = A_{sw}/sb_w$.

4.5.5 Shear stiffness K_{SP} of the encased steel profiles

Let us first consider encased sections like H or U or tubes without a concrete infill. If all the sections of the steel profiles encased in a wall are the same and if we suppose that they undergo an equal displacement under the applied external force, then the applied transverse shear $V_{Ed,SP}$, which is taken by the N profiles is shared equally between them and their angular shear deformation, γ is equal to

$$\gamma = \tau/G_s = V_{Ed,SP}/NG_sA_v$$

where A_v is the shear area of one steel profile and G_s is the shear modulus of steel (80769 MPa).

Per unit vertical length of the reference cell, the horizontal displacement $\delta_{a,ref}$ of the point of application of a unitary shear force $V=1$ applied to the N profiles is: $\delta_{a,ref} = \gamma \times 1.0 = 1/NG_sA_v$

The shear stiffness per unit vertical length for the N-encased profiles is

$$K_{SP} = NG_sA_v \tag{4.27}$$

If the N-encased steel sections are H sections with a web parallel to the wall bending axis, the shear area of one profile is the area of the two flanges: $A_v = 2Nb_f t_f$. With H sections with web perpendicular to the wall bending axis, the shear area of one profile is not less than the area of the profile web and $A_v = N\,h_w t_w$ (the root radius may be included, and then A_v is found slightly greater – see code of reference.

With circular hollow sections, the shear area of one section is $A_v = 2A/\pi$, where A is the cross-sectional area of one section.

If the encased steel sections are concrete-infilled tubes, then the contribution of concrete to shear stiffness should be taken into account. The shear stiffness K_{CFST} of N concrete-infilled tube is found as the sum of K_{SP} of the tube and of K_{infill}:

$$K_{CFST} = K_{SP} + K_{infill} \tag{4.28}$$

There is not at present a standard way to assess K_{infill}. If the infills in the N tubes are made of fibre-reinforced concrete or if they are reinforced by bars, their total shear stiffness may be assessed as the one of an uncracked concrete circular section:

$$K_{infill} = 0.9 \times 0.4\ NE_cA_{c,infill} = 0.36\ NE_cA_{c,infill} \tag{4.29}$$

If the infill is unreinforced concrete, K_{infill} may be taken as one-half of an uncracked section:

$$K_{infill} = 0.9 \times 0.5 \times 0.4\ NE_cA_{c,infill} = 0.18\ NE_cA_{c,infill} \tag{4.30}$$

4.5.6 Longitudinal shear action effects at the concrete-profile interfaces in hybrid walls

The nodes of the hybrid truss model are in the chords; they are the convergence points of the compression strut force F_{diag}, the tie force in the stirrups $F_{stirrup}$ and a vertical force $V_{Ed,l}$ which equilibrates the vertical component of F_{diag} (Figure 4.18). $V_{Ed,l}$ is transmitted in the steel components of the chord through longitudinal shear. The horizontal component of F_{diag} is a compression force which equilibrates $F_{Ed,stirrup}$ in the "reference cell" of wall (Figure 4.14).

$$F_{Ed,stirrup} = F_{comp} = F_{diag} \sin \theta = V_{Ed,RC} \qquad (4.31)$$

Over the height of $z \cot \theta$ of the "reference cell":

$$V_{Ed,l} = F_{diag} \cos \theta = V_{Ed,RC} \cot \theta \qquad (4.32)$$

For $\theta = 35°$: $V_{Ed,l} = 1.43 V_{Ed,RC}$

The resisting section of the chord includes an area A_{bars} of classical bars and the area A_{prof} of the external steel profile; on the side of the truss which is in compression due to bending, it also includes an area of concrete. On the tension side, the most critical for what concerns the longitudinal shear force $V_{Ed,l,a}$ at the profile-concrete interface, $V_{Ed,l}$ is distributed between the bars and the profile in proportion to their section area A_{prof} and A_{bars}; the shear force at the profile-concrete interface over the height of $z \cot \theta$ of the "reference cell" is

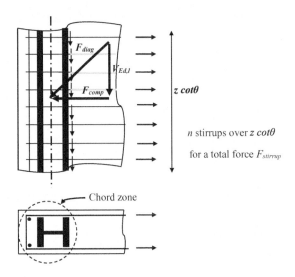

Figure 4.18 Horizontal equilibrium at a node of the truss model

$$V_{Ed,l,a} = \left(V_{Ed,l} \times A_{prof} \right) / \left(A_{prof} + A_{bars} \right) \qquad (4.33)$$

That applied shear force effect $V_{Ed,l,a}$ is distributed around the encased external profile and resisted by a combination of resistance due to shear connectors, bond and friction. Resistance by friction results from the compression forces F_{comp} and $F_{stirrups}$ applied respectively on the interior and exterior sides of the profiles.

Experiments have shown that shear connectors could be installed on the side of the profile facing the compression diagonals. For a partially encased external steel profile like in Figure 4.19 a, the total longitudinal shear $V_{Ed,l,a}$ is effectively applied on that side. Because of their distance from the chords, it may be considered that internal profiles do not participate in the truss model. As explained in Section 4.5.3, the diagonal compression force passes through and around the internal profiles (see Figure 4.19 b), and a longitudinal shear force $V_{Ed,l,a}$ is applied on each side of the profile so that a resistance to that shear force $V_{Ed,l,a}$ is necessary on both sides of each internal profile.

4.5.7 Longitudinal shear by the beam method and by the hybrid truss method

With the hybrid truss method, there is only one value of the longitudinal shear force at the steel profiles-concrete interfaces because steel sections and concrete are submitted to a single compression strut force. With the theory of beams, the longitudinal shear in external and internal profiles is different because the first

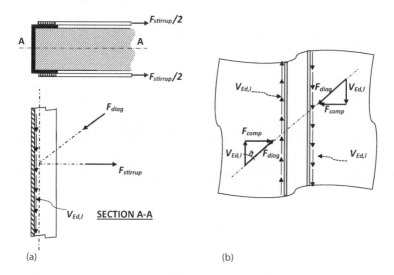

(a)   (b)

Figure 4.19 a, b At left: node equilibrium in case of a partially encased steel profile and external ties. At right, action effects at an internal profile

moment of area S is different for different profile locations. The most accurate method is the one which corresponds to the hypothesis of the method: cracked or uncracked section, which, as explained in Section 4.3, depends on the loading case. In the example of a wall submitted to a large bending moment given in Section 4.11.2, the longitudinal shear calculated by the hybrid truss method is 50% smaller than those found with the Euler-Bernoulli elastic beam method. This result has logic: the classical beam method ignores the deformability in shear of sections and, accordingly, the contribution of the encased steel profiles to the shear stiffness of the section. Therefore, with the classical beam method, the transverse shear V_{Ed} is totally taken by what is considered a classical reinforced concrete beam, while, with the hybrid truss model, only a part $(V_{Ed,RC})$ of the transverse shear is taken by reinforced concrete and the rest $(V_{Ed,SP})$ is taken by the encased profiles.

4.5.8 Contribution of encased steel profiles to the shear stiffness of walls

Steel profiles contribute to the shear stiffness of hybrid walls by the term K_{SP} additional to K_{RC} – see Section 4.5.1. In the set of test specimens summarized in Table 4.1, the ratio K_{SP}/K_{RC} for walls ranges from 0.3 to 1.8. As expected, K_{SP}/K_{RC} is, like the parameter δ_a of Eurocode 4 (2004) (see definition in Section 4.10), is greater for hybrid columns (Bogdan et al., 2019). The design generally gives some possibility to achieve large values of K_{SP} and reduce the flexibility of walls. This depends on the shear area a_s of the steel profiles, which, to some extent, is a designer's choice: with H profiles, a_s may be maximized by orienting the sections in a way such that a_s is realized by the flanges of the profiles rather than by their web.

Taking advantage of the reduced flexibility of hybrid walls due to shear stiffness requires a numerical model or a code reference which separates the term of deformation related to shear from the one related to bending. For seismic design, ACI 318-19 indicates $k_{eff}=0.35E_cI_g$ for components near the onset of yielding to be considered together with a shear stiffness of 0.2 A_gE_c. The contribution of encased sections to wall stiffness can be taken into account by multiplying the term $0.2A_gE_c$ for shear stiffness by the ratio $(K_{SP}+K_{RC})/K_{RC}$.

In Eurocode 8 (2022), the contribution of encased sections may similarly be taken into account in the calculation of the chord rotation at yield given by Equation (4.10) by multiplying the term $0.011(1+l_w/(3L_v)$ which takes into account shear deformations by the ratio $K_{RC}/(K_{SP}+K_{RC})$.

4.5.9 Shear action effects in hybrid columns

To calculate the shear action effects in hybrid columns, the section should be, as explained in Section 4.4.1 for the elastic beam method, divided into subsections. The applied global shear V_{Ed} is also distributed in the subsections

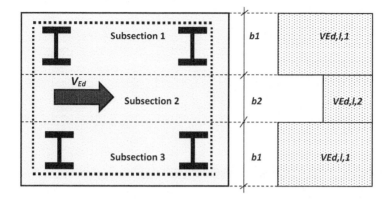

Figure 4.20 Distribution of transverse shear V_{Ed} in the subsections of a column in the hybrid truss method

in proportion to their bending stiffness $EI_{eff,i}$, applying Equation (4.12). However, the subsections should here be reinforced concrete subsections and hybrid subsections; they are thus less numerous than in the elastic beam method: three in Figure 4.20 instead of five in Figure 4.10.

4.6 RESISTANCE OF HYBRID COMPONENTS TO COMPRESSION AND BENDING

4.6.1 Resistance to pure compression

Under pure axial compression, the strains are the same in the steel and concrete parts of a composite section and the global behaviour of the section is related in a simple way to the two material laws. Plane sections remain plane under load and there is no longitudinal shear at the interfaces between the components of the component. The available compressive strength for design is a sum of the plastic strength of the components of the section (Figure 4.21), which, in the simplified method of design of Eurocode 4 (2004), may be calculated with Equation (4.34):

$$N_{pl,Rd} = A_a f_{yd} + 0.85 A_c f_{cd} + A_s f_{sd} \qquad (4.34)$$

where A_a, A_c and A_s are the area of, respectively, the steel profiles, the concrete and the reinforcement and where f_{yd}, f_{cd} and f_{sd} are the respective design strength of those components. f_{cd} is based on the cylinder strength of concrete. The constant factor 0.85 is introduced to allow for concrete strength reduction due to shrinkage, to deterioration due to environmental exposure and to splitting.

The addition in Equation (4.34) implies that the yield stress of steel and the maximum strength of concrete be reached for approximately the same

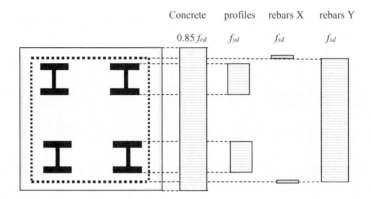

Figure 4.21 Design stresses in the components of a hybrid section in pure compression

applied strain, which is realized if steel profiles are in the range of S355 to S460, if the steel reinforcement grade is up to S500 and the concrete grade is in the range of C20 to C50. Indeed, the nominal yield strain ε_y of S355 steel is 0.0017, and its average yield strain, taking into account the material overstrength of production, is around $\varepsilon_y = 1.25 \times 0.0017 = 0.0021$; the strain ε_{c1} the compressive strain in concrete at the peak stress f_c is usually taken equal to 0.0020; the nominal yield strain ε_y of S500 steel is 0.0024 and the average yield strain is around 0.0028. All these strains are in the same range and Equation (4.34) is applicable. It remains applicable if variations around the mentioned properties are small because safety factors give a margin of variation: for instance S460 steel profiles, with nominal $\varepsilon_y = 0.0022$ and average $\varepsilon_y = 0.0022 \times 1.2 = 0.0026$, remain in the correct range.

But Equation (4.34) would not be valid for S800 steel because its nominal yield stress $\varepsilon_y = 0.0038$ is greater than $\varepsilon_{cu,2} = 0.0035$, the ultimate compressive strain in confined concrete: concrete would be crushed for an applied axial force $N_{E,d}$ smaller than the resistance calculated with Equation (4.34).

The addition in Equation (4.34) also implies that local buckling of steel does not take place before the peak stress is reached in concrete. For steel sections fully encased in concrete, provided the code requirements on the layout of longitudinal and transverse reinforcing bars and on the minimum thickness of concrete cover are met, local buckling will not occur before reaching the design resistance $N_{pl,Rd}$. For partially encased I or H sections, a limit of flange in relation to slenderness is imposed in design codes and almost all rolled sections satisfy that condition. In Eurocode 4 (2004), the limit of flange in relation to slenderness c/t_f for profiles of section class 1 is $20 \times (235/f_y)^{0,5}$.

Equation (4.34) ignores the effect of concrete confinement, which can increase the maximum strength and the strain at maximum strength at certain conditions. A confining effect can only result from the limitation of transverse deformations of concrete by the steel encasement. The confining effect can only exist when the Poisson's ratio ν of the two materials becomes significantly different – otherwise both steel and concrete undergo similar lateral expansion under longitudinal compression – and if the steel encasement is stiff and remains in the elastic range under the pressure of laterally expanding concrete. In the elastic stage of the two materials, their Poisson ratios are similar, about 0.2 for concrete and 0.3 for steel. But as concrete enters into the plastic range, its Poisson ratio ν increases progressively up to 0.5. With encased open sections like I or H, a confining effect is only due to stirrups and is negligible. The confining effect can be significant in concrete-filled sections if the concrete filler strength is not too great so that its plastic expansion takes place before the filled tube yields (Nethercot, 2004). Eurocode 4 (2004) allows a simple approach to take into account the presence of a concrete infill within the limits of material properties recalled above: the application of Equation (4.34) without the 0.85 coefficient:

$$N_{pl,Rd} = A_a f_{yd} + A_c f_{cd} + A_s f_{sd} \tag{4.35}$$

More complete approaches take into consideration all factors. When steel is restraining concrete, the tensile stress which results in the circumferential direction of the steel hollow section reduces the available yield strength f_y in the longitudinal direction, which limits the global beneficial effect of taking into account concrete confinement. This can be calculated considering the criteria of plasticity for steel and the influence of confinement on concrete strength. Design equations to estimate the maximum strength of CFST columns have been proposed (Lai et al., 2017). They result from a calibration of more than 100 experimental results and cover a wide range of material strengths and diameter-to-thickness ratios. They can be used to assess the resistance to compression of CFST boundary zones of walls. An empirical equation expresses the hoop stresses f_{s0} in the function of the unconfined concrete strength f_c and the tube diameter D, thickness t and uni-axial yield stress $f_{y,a}$:

$$f_{s0} = 0.2 f_{y,a} \left(D/t \right)^{0.35} \left(f_c/f_{y,a} \right)^{0.45} \tag{4.36a}$$

The available axial yield stress $f_{y,red}$ in the tube may be found through the application of the Von Mises criterion for a bi-axial stress state:

$$f_{y,red}^2 - f_{s0} f_{y,red} + f_{s0}^2 = f_{y,a}^2 \tag{4.36b}$$

The confined concrete stress $f_{c,c}$ at the maximum load depends on the confining stress $f_{r,a}$ due to the tube, which may be calculated with Equations (4.36c) and (4.36d):

$$f_{r,a} = - \left(2t/(D-2t) \right) f_{s0} \tag{4.36c}$$

$$f_{c,c} = f_c\left[1 + 3.5\left(\frac{f_{r,a}}{f_c}\right)\right] \tag{4.36d}$$

The design ultimate load-carrying capacity of the boundary zone infilled tube is the sum of the components of the CFST:

$$N_{Rd} = A_a f_{y,red} + 0.85A_c f_{cd} + A_{c,i} f_{c,c} + A_s f_{sd} \tag{4.37}$$

where A_a, A_c, $A_{c,i}$ and A_s are the area of, respectively, the steel tube, the unconfined concrete outside of the steel tube, the concrete infill and the reinforcement; f_{cd} and f_{sd} are the design strength of the unconfined concrete outside of the steel tube and of the reinforcement.

4.6.2 General method for assessing resistance to compression and bending

The standard procedure to calculate the resistance of hybrid components to a combination of axial compression N_{Ed} and bending moment M_{Ed} is based on the assumption that plane sections remain plane so that the strain distribution in the composite cross-section is linear.

The sequence of calculations is the following. The concrete strain in the exterior compression fibre is set to the concrete crushing strain (Figure 4.22a), and an arbitrary position for the neutral axis NA is assumed. As the strain distribution is linear, strains in the cross-section

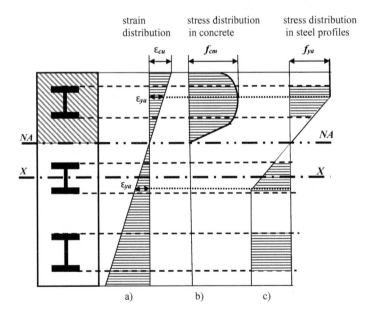

Figure 4.22 Determination of the N-M interaction diagram of a composite section

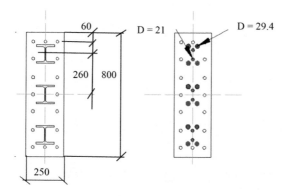

Figure 4.23 Wall with three encased HEB120 profiles. At left, the real section. At right, a model with bars of diameter D=21 mm for the web and D=29.4 mm for the flanges

are determined (Figure 4.22a). The stresses in the different components of the section are deduced from the distribution of strains and the stress-strain relationships of the constituent materials. Concrete is assumed to have no resistance to tension (Figures 4.23b). The axial resistance N_{Rd} is obtained as the difference between the resultant compression force and the resultant tension force or, equivalent, as the integration of stress over the whole composite cross-section. The bending moment resistance M_{Rd} is obtained by taking moments about the plastic centroid of the cross-section.

The general approach described above may be time-consuming unless it is carried out by software dedicated to the calculation of composite or reinforced concrete sections, like HBCOL, see 3.5. It is in general a fibre model: the section is represented by fibres, each obeying the stress-strain law of its constitutive material. Modelling a steel profile by means of two circular bars for each flange and two for the web as shown in Figure 4.23 was proved accurate (Bogdan and Chrzanowski, 2018): the N_{Rd}–M_{Rd} inter-action curves for bending about axis X-X are identical with profiles modelled explicitly or by that set of bars. This is not surprising because, if plane sections remain plane as supposed, the strain in any fibre of material only depends on its distance to the neutral axis, not on the fact that some fibres are in reality grouped in a steel profile.

For bending around axis Y-Y, a model in which a flange is represented by two bars does not figure correctly the elastic bending modulus, but the plastic bending modulus is correct if a bar representing a half flange is positioned at the centre of this half flange: the plastic resultant of each of the two bars equivalent to a half flange is at the same distance from the axis Y-Y as the half flange which it represents.

4.6.3 Simplified method for assessing resistance to compression and bending

Simplified calculations "by hand" of the $N_{Rd}–M_{Rd}$ interaction of reinforced concrete or composite sections are possible. Roik and Bergmann (1992) have proposed a simple method to determine an approximate interaction curve $N_{Rd}–M_{Rd}$ for symmetrical composite components. The method, to which reference is made in AISC360-16 and in Eurocode 4 (2004), is based on several simplifications:

- Only key points A, B, C and D of the $N_{Rd}–M_{Rd}$ curve are determined, and the curve is constructed by joining them by straight lines (Figure 4.24).
- To evaluate these key points, material behaviour is assumed rigid-plastic; steel is considered to have reached yield in either tension or compression; concrete is assumed to have reached its peak stress in compression and its tensile strength is zero.
- The concrete stress distribution is replaced by a uniform stress block over the area of concrete under consideration, but with a reduced peak compression stress.

The method is accurate enough to be used in design and gives results which are on the safe side and within the limitations of strains and the consideration for concrete-filled steel tubes explained in Section 4.6.1. Otherwise, the general approach in Section 4.6.2 should be used.

The coordinates of key points A, B, C and D and of the position of the plastic neutral axis (PNA) are calculated with simple expressions. Detailed explanations about their logic are given in Nethercot (2004). This logic is applicable to any type of composite or hybrid symmetrical section. Only the practical expressions are given hereunder. The coordinates of key points A, B, C and D in a $N_{Rd}–M_{Rd}$ diagram can be calculated as follows.

Point A $(N_{pl,Rd}, 0)$ corresponds to the pure axial compression capacity $N_{pl,Rd}$, which can be calculated with Equations (4.34), (4.35) or (4.40) (Figure 4.25).

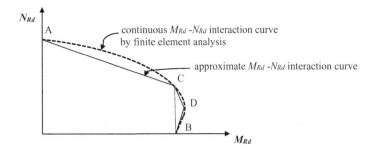

Figure 4.24 Approximate $N_{Rd} – M_{Rd}$ interaction curve with key points A, B, C and D

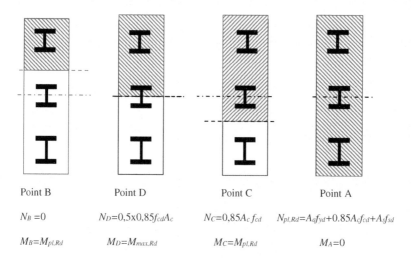

Point B	Point D	Point C	Point A
$N_B = 0$	$N_D = 0,5 \times 0,85 f_{cd} A_c$	$N_C = 0,85 A_c f_{cd}$	$N_{pl,Rd} = A_a f_{yd} + 0.85 A_c f_{cd} + A_s f_{sd}$
$M_B = M_{pl,Rd}$	$M_D = M_{max,Rd}$	$M_C = M_{pl,Rd}$	$M_A = 0$

Figure 4.25 Definition of the stress states corresponding to points A, B, C and D of the N-M interaction curve shown at Figure 4.24

Point B $(0, M_{pl,Rd})$ corresponds to the pure bending moment capacity $M_{pl,Rd}$. It requires calculating the position of the PNA. A tentative position of the PNA is defined at a distance X to a chosen reference line, for instance the axis of symmetry of the section. Due to the fact that there is more capacity for resistance in compression than in tension, since the resistance of concrete in tension is supposed to be 0, the PNA is on the part of the section in compression (Figure 4.25). The components of the plastic resistance on the compression side are calculated; some of them, for instance the area of concrete above the PNA, are a function of X. The same is done on the tension side and the plastic resistance in tension calculated; by expressing the equilibrium of the compression and tension forces, the position X of the PNA is found. The ultimate moment of resistance $M_{pl,Rd}$ is calculated by adding the moments of the compression and tension forces about any axis parallel to the PNA.

Point C $(N_C, M_{pl,Rd})$ is defined by $M_{pl,Rd}$ and $N_C = N_{c,Rd} = 0.85 \, A_c \, f_{cd}$

$N_{c,Rd}$ is the resistance to compression of the entire concrete part A_c of the composite cross-section. The PNA corresponding to Point C is symmetrical to the one of Point B (Figure 4.25).

Point D (N_D, M_{max}) is the maximum bending moment point; it is given by:

$$N_D = 0.5 \times 0.85 \, A_c \, f_{cd} \quad M_{max,Rd} = W_{pa} f_y + 0.5 \, W_{pc} f_{ck} + W_{ps} f_y \quad (4.38a)$$

where W_{pa}, W_{pc} and W_{ps} are the plastic moduli of, respectively, the steel profiles, the overall concrete and the reinforcement about the axis of symmetry of the composite cross-section. Note that alternatively to the way explained above, $M_{pl,Rd}$ can be calculated as:

$$M_{pl,Rd} = M_{max,Rd} - (W_{pan}\,f_y + W_{pcn}\,f_{ck} + W_{psn}\,f_{sk}) \qquad (4.38b)$$

where W_{pan}, W_{pcn}, and $W_{psn}\,f_{sk}$ are the plastic moduli of, respectively, the steel profiles, the concrete and the reinforcement within a central zone of the section limited by +X and –X ordinates.

The PNA corresponding to Point D is at the axis of symmetry of the section (Figure 4.25).

In the calculations of the characteristics of the hybrid section, simplifications are possible.

As explained in Section 4.6.2, in a rigid-plastic approach, a steel profile may be modelled as a single bar rather than by explicit geometry of flanges and web, but this representation may be inadequate if the PNA passes through the profile; in such a case, specific care should be taken in the definition of the equivalent bar: it should be at least a rectangle of height equal to the one of the profile so that the portion of profile or equivalent bar which are above or below the PNA is the same. To that end, H sections should be modelled by rectangular bars which have the following properties:

- A height d equal to the depth d of the section:
- A width $b^* = A_a / d$ where A_a is the area of the H profile section.

That part of the equivalent rectangle above the PNA does not correspond exactly to the part of the profile, but that approximation is acceptable, as it concerns a minor contribution to the overall section resistance.

Concerning classical reinforcement, calculations can become time-consuming if there is a great number of longitudinal bars and if every single bar is modelled separately. The process may be made more user-friendly by replacing a line of n rebars with an equivalent plate such that its plastic strength in tension and bending are the same as those of the set of rebars that it represents (Figure 4.26). The characteristics of a plate equivalent to n rebars are:

- Plate area $A_p = n\,A_b$, where A_b is the cross-sectional area of one bar.
- Plate length l_p : $l_p = d_b\,(n+1)/n$, where d_b is the distance between the two outer bars in the line.

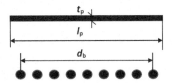

Figure 4.26 Layer of rebars and equivalent plate

- Plate thickness: $t_p = A_p / l_p$.
- For a line of bars parallel to the bending axis, the distance d_p of the equivalent plate geometrical centre to the plastic neutral axis is equal to d_p of the bars.

The assessment of the simplified method of determination of the M_{Rd}–N_{Rd} interaction curve has been made in several test programs, by comparison to either experimental results (Bogdan et al., 2019) or to numerical results obtained with fibre models. They conclude with the adequacy of the simplified method, which is on the safe side with ann approximation within a few percent of the real strength.

An example in Section 4.11.1 illustrates the simplified method of calculation of the design strength of a wall with several embedded steel profiles of the type shown in Figure 4.27 under a combination of axial force N_{Ed} and bending moment M_{Ed}. For hybrid cross-sections of the types shown in Figures 4.27a and b, relationships of the N_{Rd}–M_{Rd} interaction curves may be found in the literature (Plumier et al., 2012; Gerardy et al., 2013; Trabuco et al., 2016; Bogdan et al., 2019).

4.6.4 Reduction of bending resistance by shear stresses

Shear can reduce the resistance of profiles to tension and to compression (also to bending but this is not of interest for profiles used as reinforcement in walls or columns).

In Eurocode 3 (2021), it is indicated that the effect of shear on the resistance of steel profiles to tension or compression may be taken as negligible if the applied transverse shear V_{Ed} in a profile is smaller than one-half of its plastic resistance to shear $V_{pl,Rd}$:

$$V_{Ed} \leq 0.5\ V_{pl,Rd} \tag{4.39}$$

If the applied shear V_{Ed} is greater than $0.5\ V_{pl,Rd}$, the tension or compression resistance of the cross-sectional area A_v affected by shear is reduced. The resistance $N_{pl,Rd}$ of profiles to tension and compression becomes:

$$N_{pl,Rd} = (A - \rho A_v)\ f_{yd} \tag{4.40a}$$

where: $\rho = (2V_{Ed}/V_{pl,Rd} - 1)^2$ \tag{4.40b}

This reduction in strength, like the cross-sectional area A_v affected by shear, depends on the orientation of the profiles relative to the shear force direction.

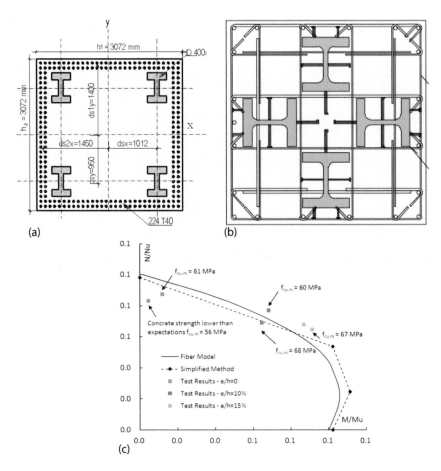

Figure 4.27 a, b, c Top: examples of hybrid column sections for which calculated examples of M-N interaction curve are available. Bottom: test results for the section at top right and M_{Rd}-N_{Rd} interaction curves calculated by F.E. analysis and by the simplified method (Bogdan et al., 2019)

4.7 RESISTANCE OF HYBRID COMPONENTS TO SHEAR

4.7.1 Resistance to transverse shear

AISC360-16 mentions three possibilities to calculate the available shear strength of a reinforced concrete component with encased composite sections:

1) The shear strength of the steel sections alone.
2) The shear strength of the reinforced concrete alone.
3) The addition of the shear strength of the steel sections and of reinforced concrete.

These different possibilities do not possess the same degree of accuracy in estimating the real available shear strength. Consideration of the only shear resistance of the steel profiles as proposed in 1) certainly does not correspond to the real behaviour in the shear of column or wall sections which are essentially reinforced concrete. Disregarding the contribution to the shear resistance of the net area A_c of concrete would clearly be a gross underestimation for a large section with several encased steel profiles. Possibility 2) is also on the safe side, as it ignores the contribution of steel profiles to shear strength, but it corresponds better than possibility 1) to the reality of columns or walls which are essentially reinforced concrete sections. However, possibility 2) requires two adaptations: first, the section considered for shear resistance should be the gross concrete section, like in reinforced concrete, steel profiles being just another type of rebar; no one would think of checking reinforced concrete sections' resistance to shear taking out the rebar section area; second, as shown by Figure 4.10, the distribution of shear in the section is neither uniform nor regular and global section checks performed like for a standard reinforced concrete section would ignore this fact. Care should be taken to identify and check the most stressed zones, which are the hybrid subsections. To be effective, transverse reinforcement should follow the distribution of transverse shear in the section. In particular, the transverse steel reinforcement of the hybrid subsections should be placed around these subsections.

The design possibility 3) is attractive in that it should give the most accurate estimate of the real shear strength of hybrid sections, but the way to check hybrid subsections in shear is not standard. Hybrid subsections are chains of concrete and steel components and their strength may be one of the brittle links, the concrete links. With design checks expressed in terms of shear stresses in the frame of the classical beam theory, which is valid with sections made of materials working in the elastic range, the only practical way to be on the safe side consists in considering that hybrid subsections are reinforced concrete subsections and apply on them the classical checks of reinforced concrete. In fact, this means the use of design possibility 2). For hybrid components, design possibility 3) requires another reference model, which can be the hybrid truss model explained in Section 4.5.

4.7.2 Resistance to longitudinal shear at the steel-concrete interface

At a steel-concrete interface, the resistance $V_{Rd,l,a}$ to an applied longitudinal shear $V_{Ed,l,a}$ can be provided by bond, friction and shear connectors, with appropriate partial safety factors being considered. As explained in Section 4.5.6, the design check should be respectively:

- In one external profile: $V_{Ed,l} \leq V_{Rd,l}$ (4.41a)
- In one internal profile: $2V_{Ed,l} \leq V_{Rd,l}$ (4.41b)

Eurocode 4 (2004) allows to sum up the bond, friction and shear connectors contributions in order to obtain the necessary total resistance $V_{Rd,l}$:

$$V_{Rd,l} = V_{Rd,bond} + V_{Rd,friction} + V_{Rd,connectors} \tag{4.42}$$

The resistance to longitudinal shear due to bond $V_{Rd,bond}$ is found as the product of the design shear strength τ_{Rd} given in Table 2.2 times the height $z cot\theta$ of the "reference cell" (see Section 4.5.2) times the perimeter of the steel profile:

$$V_{Rd,bond} = \tau_{Rd} \times z\ cot\theta \times l_{profile,perim} \tag{4.43a}$$

In assessing the experimental results of tests on hybrid walls, bonds may be considered active on the complete perimeter of the section. However, in design, there may be doubt on the surface state of rolled profiles and one may wish to manage the uncertainties related to bond strength, for instance by considering only one-half of the profile perimeter $l_{profile,perim}$ as active. If so:

$$V_{Rd,bond} = 0.5 \times \tau_{Rd} \times z\ cot\theta \times l_{profile,perim} \tag{4.43b}$$

Friction forces $V_{Rd,friction}$ resisting longitudinal shear are the result of compression struts taking reaction on the steel profiles, which, as explained in Section 4.4.6, apply on them a normal force $V_{Ed,RC}$:

$$V_{Rd,friction} = \mu\ V_{Ed,RC} \tag{4.43c}$$

where μ is the friction coefficient at a steel-concrete interface.

There may be other friction forces, for instance due to external forces applied on the side of the wall, in its plane, or by a reaction at a support. This situation is more likely with hybrid beams; for walls it would correspond to unusual architectural design.

Resistance $V_{Rd,connectors}$ to longitudinal shear can be provided by shear connectors like headed studs or welded plates – see Chapter 2 for the design of these components.

Plates welded on an encased profile, as in Figure 2.15, can realize a direct bearing for the concrete compression struts and provide resistance to longitudinal shear. They can be designed by a "strut-and-tie" method. A design in which compression struts head in direction of the wall section height rather than width is preferable because it avoids compression struts creating an outward thrust which must be resisted by additional stirrups or links to prevent spalling of the concrete and loss of strength of the component. If, however, an outward thrust has to be resisted, the stirrups or links should be present at each connector and designed to resist a tension force equal to the projection on a horizontal of the shear force capacity of the welded plate connector.

4.7.3 Resistance of hybrid walls to transverse shear in static condition

As explained in Section 4.7.1, there are three possibilities to determine the available shear strength of a component with encased composite sections. Briefly said, they are: steel alone, reinforced concrete alone, and the sum of steel and reinforced concrete.

In the frame of the classical beam method, the contribution of steel profiles and reinforced concrete to shear stiffness is not calculated, so the distribution of the applied shear action effect between steel profiles and concrete is not known. The check of resistance may be made considering that the necessary resistance to transverse shear is provided by the reinforced concrete alone, with design resistance V_{Rd} greater than design shear V_{Ed}. In Eurocode 2 (2004), two design resistances V_{Rd} to shear are considered for reinforced concrete, each resistance corresponding to a different ultimate limit state: $V_{Rd,max}$ corresponds to the crushing of concrete compression struts; $V_{Rd,s}$ corresponds to yielding of the transverse reinforcement, supposed horizontal in walls or columns; the shear resistance, V_{Rd} should be taken as the lesser of $V_{Rd,max}$ and $V_{Rd,s}$. The design checks are:

$$V_{Ed,RC} \leq V_{Rd,max} = \alpha_{cw} b_w z \, \nu_1 f_{cd} \, / (cot\theta + tan\theta) \qquad (4.44)$$

$$V_{Ed,RC} \leq V_{Rd,s} = \frac{A_{sw}}{s} z f_{ywd} cot\theta \qquad (4.45)$$

with the coefficients ν_1 and α_{cw} explained in Section 4.7.5.

However, in AISC360-16, it is explicitly stated that the shear resistance may be considered as the sum of the plastic strength in the shear of steel profiles and of the transverse steel reinforcement, a second term equivalent to $V_{Rd,s}$ of Eurocode 2 (2004). In static conditions, this is valid without requiring a hierarchy between $V_{Rd,s}$ and $V_{Rd,max}$, which, as explained in Section 4.7.4, may be necessary when ductility is a design objective additional to strength.

With the hybrid truss model, the distribution of the shear action effect between steel profiles and concrete is calculated, and it is possible to make separate checks for reinforced concrete and for steel profiles and to take advantage, on the reinforced concrete side, of a design action effect $V_{Ed,RC}$ which is smaller than V_{Ed}. Then the design checks are:

$$V_{Ed,RC} \leq V_{Rd,max} \qquad\qquad V_{Ed,RC} \leq V_{Rd,s}$$

$$V_{Ed,SP} \leq V_{pl,Rd,tot} = N \, V_{pl,Rd} \qquad (4.46)$$

where $V_{pl,Rd}$ and $V_{pl,Rd,tot}$ are respectively the design resistance to shear of one encased profile and of the set of N encased profiles.

4.7.4 For a ductile behaviour of hybrid walls submitted to bending and transverse shear

In earthquake-resistant design, it is often necessary to create conditions for the ductile behaviour of columns or walls. Compliance with the design conditions in Section 4.7.3 provides strength, but not necessarily ductility.

Yielding of flexural components can lead to degradation of the shear-resisting mechanisms. To realize the ductile behaviour of hybrid walls submitted to bending and transverse shear, a hierarchy of failure modes should be realized, such that ductile failure modes take place first; for that reason, they should be the ones with the smaller resistance.

The most ductile failure mode in reinforced concrete is flexural. The failure mode in shear by crushing compression struts in concrete provides almost no ductility. The failure mode in shear related to plastic elongation of transverse reinforcement provides some ductility, but, under progressively increased cyclic displacements, there is a degradation of resistance to shear consecutive of the progressive elongation of transverse reinforcement and of the progressive cracking. Figure 4.28, which shows load-displacement relations of walls with confined boundary elements, illustrates how design for shear influences this progressive degradation of resistance: wall B4, designed to yield in flexure and submitted to monotonic loading, exhibits a very ductile behaviour without significant loss of resistance up to a 7% drift ratio; for wall B3, similar to B4 but submitted to steps of cyclic loading, this drift ratio decreases to about 4.5%; for wall B5, similar to B4 but less slender and in which there is no overstrength in shear resistance, fails by crushing of diagonal compression struts in the web and the drift ratio capacity is only 2.8%.

To take advantage of the plastic character of shear resistance related to the yielding of reinforcements together with the inevitable strength degradation with greater imposed displacement, a margin of safety has to be

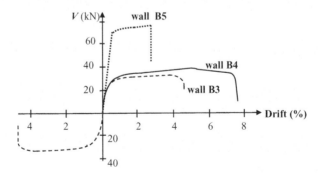

Figure 4.28 Load-displacement relations of walls with confined boundary elements illustrating the influence of shear on the progressive degradation of resistance; wall B4 is submitted to a monotonic loading; walls B3 and B5 to cyclic loading; the curves are envelope of experimental results in (Corley et al., 1981)

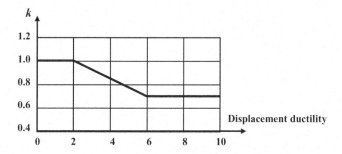

Figure 4.29 Relation between the shear strength degradation parameter k and the displacement ductility of columns based on Equation (6.4) in ASCE41-06, 6.4.2.3.1

taken in the design for shear resistance. Several empirical models expressing shear strength degradation of flexural components have been developed which may help in this perspective. The model by Sezen and Moehle (2004), which is reported in ASCE 41, reflects the shear strength degradation as a function of the displacement ductility by means of a parameter k shown in Figure 4.29. Depending on the displacement ductility required in a project, the diagram of Figure 4.29 can help determine how much $V_{Rd,max}$ should be greater than $V_{Rd,s}$ in order to exploit some ductility in shear: if a great local ductility is envisaged because large force reduction factors are used a design, k should be taken equal to 0.7, and the condition on $V_{Rd,max}$ would be: $V_{Rd,max} \geq V_{Rd,s}/0.7 \approx 1.5\ V_{Rd,s}$.

Such a condition is present in seismic design codes, under various presentations. In the revised Eurocode 8 (2022), the factor ν which serves to define the ratio between concrete strut strength and standard concrete strength in static design is, for the design of ductile structures, divided by 1.6.

Based on the explanations above, the proposed design conditions for the ductility of hybrid walls or columns are:

$$M_{Rd}\ (N_{Ed}, V_{Ed}) \geq M_{Ed} \tag{4.47a}$$

with $M_{Rd}\ (N_{Ed}, V_{Ed})$ as close as possible to M_{Ed} : $M_{Rd}\ (N_{Ed}, V_{Ed}) \approx M_{Ed}$

$$V_{Rd,s} \geq \gamma_{V1}\ V_{Ed,RC} \tag{4.47b}$$

where γ_{V1} = 1.0 for low ductility (displacement ductility below 2)
γ_{V1} = 1.2 for medium ductility (displacement ductility below 4)
γ_{V1} = 1.5 for high ductility (displacement ductility greater than 4)

$$V_{Rd,max} \geq \gamma_{V2}\ V_{Rd,s} \tag{4.47c}$$

where γ_{V2} = 1.2 for low ductility (displacement ductility below 2)

γ_{V2} = 1.4 for medium/high ductility (displacement ductility greater than 2)

In walls of low slenderness, condition (4.47a) may be impossible to realize because the resistance M_{Rd} in bending is inevitably much greater than the action effect M_{Ed} and failure in shear is the mode with the smaller resistance. Some ductility can be found by complying with conditions (4.47b) and (4.47c). The parameter defining the slenderness of walls is the shear span ratio or shear aspect ratio λ of the wall:

$$\lambda = M_{Ed} / (V_{Ed}\, l_w) = H/l_w \qquad (4.48)$$

where M_{Ed} is the applied moment, V_{Ed} the applied shear and l_w the height of the wall section. Failure in bending rather than shear requires that λ be greater than 1.5 to 2. The experiments on walls with several embedded steel profiles have shown that ductility exists if the conditions defined in the existing codes concerning the shear aspect ratio and the confinement of the boundary regions are respected – see Section 4.8.

4.7.5 Combining reinforced concrete and encased profiles for resistance and ductility

We have seen in Section 4.7.4 that respecting a hierarchy of failure modes is a condition for the ductility of reinforced concrete walls. There are three other conditions with hybrid walls; two are related to the possibility of adding the plastic shear resistance of the encased profiles and of the truss representing reinforced concrete; a third condition is related to the influence of axial forces on ductility in bending.

4.7.5.1 Conditions for the addition of resistance to shear of reinforced concrete and of profiles

Yielding in shear of steel profiles is a plastic mechanism, in fact the most ductile of all plastic mechanisms in steel structures. The plastic shear strength V_{Rd} of a hybrid wall may be taken equal to the sum of the plastic shear strength $V_{pl,Rd,tot}$ of all the encased steel sections and the plastic shear strength $V_{Rd,s}$ of the reinforced concrete truss, if two conditions are fulfilled:

- Condition 1: the plastic mechanism of the reinforced concrete truss develops in the same range of shear deformations γ as the plastic shear mechanism of the steel profiles;
- Condition 2: the two plastic mechanisms have adequate ductility.

Are these two conditions easily realized in hybrid walls? Let us consider as an example a wall with two encased S355 steel profiles as chord reinforcement and with concrete resistance f_{cd} and transverse reinforcement designed such that $V_{Rd,max} = 1.5V_{Rd,s}$.

The shear angular deformation at yield in shear of a steel profile is given by:

$$\gamma_{SP,y} = \tau_{pl,Rd} / G \Rightarrow \gamma = 355 / \left(80769 \times \sqrt{3}\right) = 0.0025 = 0.25\%$$

The ductility in shear of the steel profiles, without loss of strength (in fact with an increase of strength due to strain hardening), is at least of the order of 12, as indicated in Tables 4.5 and 4.6 of FEMA356 (FEMA 356, 2000) for steel columns panel zones. This means that the plastic resistance $V_{pl,Rd,tot}$ of profiles in shear is available from a drift equal to 0.25% up to a drift equal to, at least: $\gamma=12 \times 0.0025 = 0.03$ or 3%.

Considering concrete compression struts inclined at 45°, the resistance of a wall in shear, if yielding of transverse steel reinforcement governs, is

$$V_{Rd,s} = \frac{A_{sw}}{s} z f_{ywd}$$

with, for S500 reinforcement: $f_{ywd} = 500/1.15=435$MPa. If crushing of concrete compression struts governs, the resistance to shear of a wall made of C40 concrete ($f_{ck} = 40$ MPa; design strength $f_{cd} = 26.6$MPa, modulus $E_c = 35000$ MPa), is $V_{Rd,max} = \alpha_{cw} b_w z \nu_1 f_{cd}$.

ν_1 is a strength reduction factor for concrete cracked in shear which depends on the stress level in the reinforcement. At the yield of the reinforcement, $\nu_1 = 0.6(1 - f_{ck} / 250) = 0.5$ (with f_{ck} in MPa). α_{cw} takes account of the interaction of the stress in the compression chord with an applied axial compressive stress σ_{cp}; for instance with $0.25\,f_{cd} < \sigma_{cp} \le 0.5\,f_{cd}$: $\alpha_{cw} = 1.25$

It results in $V_{Rd,max} = \alpha_{cw} b_w z \nu_1 f_{cd} = 1.25 \times 0.5 \times b_w z f_{cd} = 0.625 \times b_w z f_{ca}$.

To achieve ductility of the reinforced concrete truss, Equation (4.47c) imposes an overstrength of $V_{Rd,max}$ over $V_{Rd,s}$: $V_{Rd,s} = 0.625 \times \dfrac{b_w z f_{cd}}{1.5} = 0.417\,b_w z f_{cd}$.

At yield of the stirrups, the displacement δ of the "reference cell" of reinforced concrete – see Section 4.5.1 – under a shear force $V_{Ed,c}$ equal to $V_{Rd,s}$ can be calculated with Equation (4.19):

$$\delta = \frac{V_{Rd,s}s}{E_s A_{sw}} + \frac{V_{Rd,s}}{0.25 E_{cm} b_w} = z \times \left(\frac{f_{ywd}}{E_s} + \frac{0.417 f_{cd}}{0.25 E_{cm}}\right) = z \times \left(\frac{415}{210000} + \frac{0.417 \times 26.6}{0.25 \times 35000}\right)$$

- $\delta = z \times (0.00197 + 0.00126) = 0.00323\,z$

The shear angular deformation at yield in shear of the reinforced concrete truss is given by:

$$\gamma_{truss,y} = \delta/z = 0.00323 = 0.32\%$$

This example sets forward features of the considered hybrid wall which are positive for ductility:

- The steel profiles yield before the reinforced concrete truss because $\gamma_{SP,y}=0.25\%$ is smaller than $\gamma_{truss,y}= 0.32\%$.
- The plastic resistance of profiles in shear $V_{pl,Rd,tot}$ is available up to $\gamma = 3\%$ which is nine times the drift $\gamma_{truss,y}= 0.32\%$ at which the stirrups start yielding.
- The addition of the plastic resistance $V_{Rd,s}$ of the reinforced concrete truss and $V_{pl,Rd,tot}$ of the steel profiles may thus be valid in a range from $\gamma = 0.25\%$ to $\gamma = 3\%$.
- If we refer to Figure 4.29, no loss of resistance would be observed in the range of γ from 0.32 % to 0.64 % and the loss of strength would be about 30% of $V_{Rd,s}$ in a range of γ from 0.64% to approximately 1.86% ($1.86 \approx 6 \times 0.32$).
- Then, the resistance would remain constant up to $\gamma = 8 \times 0.32\% = 2.5\%$.
- The numbers given hereabove correspond to a particular set of data ($\theta=45°$, S500 reinforcement, grade S355 profiles, $f_{ck}=40$ MPa), but the observed trend may easily be confirmed with other data. With grade S460 steel profiles, all other data being kept unchanged, $\gamma_{SP,y}$ and $\gamma_{truss,y}$ would both be equal to 0.32% and the steel profiles and the concrete truss would yield simultaneously. With concrete grades higher than C40, the hierarchy of $\gamma_{SP,y}$ and $\gamma_{truss,y}$ and the conclusions would be the same as in the example calculated above.
- The loss of shear strength in a hybrid wall is only on the reinforced concrete side and it affects only the term $V_{Rd,s}$ in a total shear resistance equal to $(V_{Rd,s} + V_{pl,Rd,tot})$; in a range of γ from 0.6% to approximately 2.6%, the loss of shear strength in a hybrid wall may be significantly smaller than 30%: in the design example in Section 4.11.2, the loss of strength on $(V_{Rd,s} + V_{pl,Rd,tot})$ is equal to only $0.3 \times 2247/(2247 + 6507) = 7.7\%$.
- The resistance at the significant damage or SD limit state is generally taken as the load, which corresponds to a 20% loss of resistance in comparison to the maximum observed resistance; Figure 4.29 indicates that, in reinforced concrete columns, a 20% loss of strength corresponds to a displacement ductility of 4.6. As explained above, the 30% loss of strength can become smaller than 20% if the input $V_{pl,Rd,tot}$ of steel profiles to shear resistance is greater than one-half of the concrete truss shear resistance $V_{Rd,s}$; then the available displacement ductility becomes greater than 8 ($8=2.6/0.32$).
- The previous conclusion indicates that the most ductile solution is obtained by choosing a layout of encased profiles with the larger possible shear area. If the flange area of classical H sections is currently double the web shear area, the best orientation of profiles in a wall would be with flanges parallel to the long side of the wall, like in Figure 4.17b.

- The concrete strength f_{ck} intervenes marginally, only through its relationship with the modulus E_c, thanks to the design condition by which $V_{Rd,max}$ is greater than $V_{Rd,s}$.

A transverse shear-drift diagram corresponding to the considerations and calculations above is shown in Figure 4.31. In this diagram, $V_{Rd,RC}$ and $V_{Rd,SP}$ are in proportion to those parameters in the experiment described in Degée et al. (2017).

4.7.5.2 Condition on axial force for ductility of hybrid walls in bending

Eurocode 8 (2022) defines limitations to the normalized design axial load $\nu_{d,c}$ for ductile composite walls in the seismic design situation. $\nu_{d,c}$ is defined as follows:

$$\nu_{dc} = \frac{N_{Ed}}{f_{cd}\left[A_c + n_0\left(A_s + A_a\right)\right]} \tag{4.49}$$

where A_a, A_s and A_c are respectively the total area of the encased structural steel cross-sections, of the steel reinforcement and of the concrete and n_0 is the ratio E_s/E_{cm} of steel and concrete modulus. Limitations are imposed to $\nu_{d,c}$ in order to limit the strains in concrete due to axial compression, a condition for ductility in the bending of hybrid walls. The imposed values result from experimental observations: $\nu_{d,c} \leq 0.40$ (medium ductility walls) or $\nu_{d,c} \leq 0.35$ (high ductility walls).

In other codes, like ACI318-19 or AISC360-16, the limitation is expressed directly in terms of strain in concrete.

4.7.6 Shear resistance of hybrid columns

The design for shear resistance of hybrid columns consists of separate checks of the reinforced concrete subsections and the hybrid subsections which are defined in Section 4.5.9. For reinforced concrete subsections, the design for shear resistance is made to Eurocode 2 (2004) with Equation (4.44) and (4.45); for hybrid subsections, the design is made as explained for hybrid walls in Sections 4.5 and 4.7.1 to 4.7.5.

4.8 LOAD INTRODUCTION AND ANCHORAGE ZONES

4.8.1 B regions and D regions

Two types of regions can be distinguished in structural components as explained in St Venant (1844): regions remote from local force introduction and regions close to local force introduction.

Regions sufficiently remote from local force introductions or supports are such that models based on the hypothesis of plane sections remaining plane are accurate enough to assess the fields of stresses in the component; those regions are continuity zones or *B regions* where "*B*" stands for Bernoulli or beam. The calculation of the shear action effects in hybrid components by the elastic beam model is only valid in *B regions*.

Regions close to local force introductions or supports, in which the field of stresses may be more complex, are named *D regions* where "*D*" stands for discontinuity or disturbed. These disturbances depend on the way the load introduction is realized. With a side force introduction on the width of a section, disturbances do exist up to a certain distance of the order of the section height, with all materials. The beam model is unable to represent stress fields in *D regions* of beams, in particular in reinforced concrete or hybrid walls, in which cracks may fan out of from the discontinuity points. Strut and ties models like the truss model shown in Figure 4.32 or the hybrid truss model developed in Section 4.5 can represent the stress field in a *D region* of a reinforced concrete component, but experience is needed in order to choose the inclination of the compression struts to make such models relatively accurate.

4.8.2 Introduction of horizontal forces in walls

Walls in buildings are most often loaded in their plane by horizontal forces at storey level where they merge with collector beams and/or slabs over the wall section height l_w. This distributes the forces from the beams and/or slabs along l_w, which correspond to a *B region* type of stress state there, like in regions in between the storeys. In slender walls, the compression struts are inclined at a constant angle as shown in Figure 4.33a, thus representing correctly the compression field, except at the yielding stage at the basis of walls designed for ductility in bending; there, flexural cracks close to the horizontal appear first and shear cracks usually form as inclined extensions of those flexural cracks and they initially propagate at approximately 45° of the vertical axis.

The design indications for the longitudinal and transverse reinforcement of beams or slabs are not different from those applied with classical reinforced concrete walls. In order to be able to distribute into the wall i the shear force $V_{Ed,storey,i}$ calculated in the analysis of the structure, longitudinal reinforcement of beams or slabs framing into a wall as shown in Figure 4.33 should realize a continuity of strength. Given the relative importance of the contribution of encased profiles to shear resistance, it is suggested that the design resistance of those reinforcements should not be smaller than $V_{Ed,storey,i}$ over the whole wall section length. The resistance to shear along cold joints (joints between concrete poured at different ages) should be checked; the dowel effect provided by the encased profiles normally provides the required resistance. In the plane of the slab, a strut-and-tie model may be used to calculate the slab reinforcement perpendicular to the wall section length necessary to resist the in-plane inertia forces due to earthquake or the forces due to wind.

In buildings, there is rarely a concentrated force applied on the short edge of the wall. If it is the case, a *D region* exist due to the combination of such a concentrated force with compression or tension forces of the chords due to global bending and with inclined compression struts of the shear resistance mechanism. All these components have to be taken into account in the assessment of the strength of that *D region*.

4.8.3 Introduction of vertical forces in walls

Vertical loads applied at the storey level to the encased steel profiles should be spread in the hybrid wall section by shear along the interface concrete-steel profiles. The design measures for such a force introduction are the same as those required for composite columns. The force transfer should be realized within a relatively short height of the order of the encased profile section height or width, whichever is the greater of the two. The transmission by shear can be realized in one of the different ways explained in Chapter 2.

4.8.4 Anchorage of steel profiles

Where they are stopped, steel profiles have to be anchored like any other reinforcement, remembering that, as explained in Section 4.1.3, they do not offer great resistance to longitudinal shear at their interface with concrete. This may lead to quite a large necessary anchorage length. With bare encased profiles, the resistance of the anchorage is provided only by a bond, and the anchorage length is about ten times greater with encased profiles than with bars. If the anchorage resistance is realized by means of classical shear connectors complying with rules on their inter-distance, the anchorage length is still large. It can be shortened by means of plate connectors, like at the free end of encased profiles shown in Figure 4.34, or by welding the encased steel profiles to horizontal anchor beams embedded in the foundation beam or raft as shown in Figure 4.34. Welded connections are preferable to bolted ones, in order to avoid a relative displacement between the vertical and horizontal steel components.

If a side horizontal force is applied at the level where a steel profile is interrupted, the calculation of the force to be resisted by the anchorage and of the necessary anchorage length should be based, like in reinforced concrete, on the consideration of a strut-and-tie mechanism at the point where the concentrated force or reaction is applied.

This *D region* type of situation exists at the supports in experimental set-ups simulating walls by tests on beams; it also exists near the force application point in tests on cantilever walls. It is fundamental to apply the requirements of reinforced concrete design codes which impose to realize an effective anchorage resistance which is of the order of the applied force F and which, as shown in Figure 4.34, may be provided in part by friction in the force application zone and by shear resistance further than the axis of the reaction force; if not, a relative sliding displacement Δ_{slip} between profiles and concrete takes place in the ends of beam span, which modify significantly the field of stresses in that zone and reduce the overall stiffness of the tested component. In order to avoid a significant Δ_{slip} the mentioned anchorage should be realized within a relatively short length of the order of the encased profile section height.

In tested specimens, the encased profiles which are on the side opposite to the applied force or reaction should also be anchored in order to avoid a relative sliding displacement Δ_{slip}. This is obvious if cyclic testing with load reversal is performed. Under monotonic loading, if sliding of the profile on the compression side takes place, a corner of the tested component is a "dead zone" in the strut-and-tie behaviour, which is not representative of a continuous wall covering several storey height, in which plane sections remain practically plane. Sliding of the profile on the compression side may lead to a low stiffness which is not representative of a continuous wall. To avoid misleading test results, two measures should be taken in the definition of test specimens of walls: 1) the concentrated force F should be distributed over the wall section height l_w by means of a spreader beam (enlarged wall thickness such as in Figure 4.30 and/or locally increased transverse reinforcement) and 2) the anchorage tension force F should be resisted by an effective anchorage of the encased profiles.

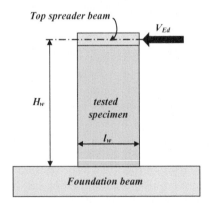

Figure 4.30 Test set-up with spreader beam and foundation distributing the applied force and the reaction force in a wall. The shear span ratio in a cantilever is H_w/l_w

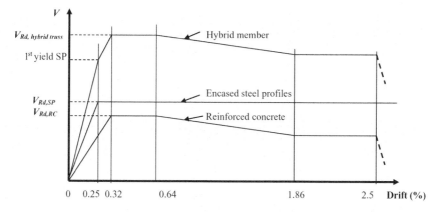

Figure 4.31 Addition of profiles and reinforced concrete shear resistance in a hybrid component

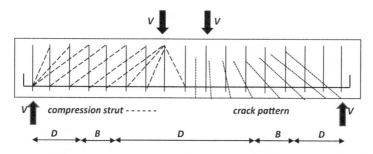

Figure 4.32 Crack pattern and compression struts in B regions and D regions in a reinforced concrete beam submitted to forces V

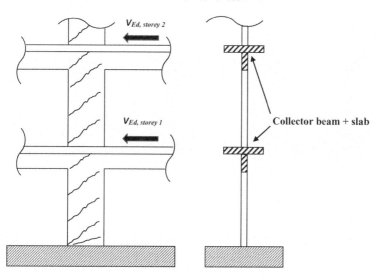

Figure 4.33 Transmission of storey shear forces $V_{Ed,storey}$ into a slender wall

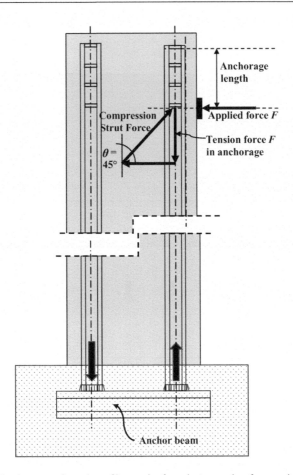

Figure 4.34 Anchorage of steel profiles at the foundation and at free end

4.9 EXPERIMENTAL RESULTS AND CALCULATIONS

4.9.1 Objectives of an overview of experimental results

After some first experimental tests on hybrid walls in 1999 (Tupper, 1999; Cho et al., 2004), more than 50 tests on hybrid walls and columns have been realized worldwide. They are described in Section 4.9.3, and their results are summarized in Table 4.1. This information and data are taken into consideration with several objectives:

- To give guidance in Section 4.10 on the limitations to longitudinal and transverse steel contents in hybrid columns or walls.
- To give a basis to physical or empirical models for the calcula-tion of the deformation and strength of hybrid columns or walls

which may be used in force-based and displacement-based analysis of structures.

- To confirm design values for longitudinal shear action effects and design resistance at encased profile interface with concrete calculated with Equations (4.33) and (4.41) to (4.43).
- To provide information on adequate layout of sections with encased steel profiles and to warn on inadequate detailing.

4.9.2 Bending stiffness $\alpha E_c I_g$ of hybrid walls or columns, from experiments to design

Values of the coefficient $\alpha E_c I_g$ for hybrid walls can be derived from experimental results. For a given specimen tested as a cantilever like in Figure 4.30 or 4.34, they are such that Equation (4.50) is satisfied. Equation (4.50) relates the applied force Q_y which generates yield in bending and the displacement δ_{exp} at the load application level under that force:

$$\delta_{exp} = Q_y \, H^3 / 3 \, \alpha_{EcIg} E_c I_g \qquad (4.50)$$

Values of α_{EcIg} for hybrid walls derived from experimental results are reported in Table 4.1. Information on the experiments to which reference is made is given in Section 4.9.2. For each given series of tests, the values in Table 4.1 are averages for the walls that comply with the design recommendations given in this chapter; results obtained with detailing, possibly bad, tested for research purposes and which did not provide a satisfactory strength or ductility are not reflected in Table 4.1. In the table, λ is the shear span ratio and ν_{exp} is the ratio of the axial force N_{exp} applied in a test to the plastic resistance of the hybrid section, without the factor 0.85 for the long-term effects of Equation (4.34):

$$\nu_{exp} = \frac{N_{exp}}{f_{cd} A_c + f_{ys} A_s + f_{ya} A_a}$$

where A_a, A_s and A_c are respectively the total area of the structural steel cross-sections, of the steel reinforcement and of the concrete, f_{ys} and f_{ya} are respectively the actual yield strength of the encased profiles and of the bars, and f_{cd} the design strength of concrete. ρ_t, ρ_l and δ_a are respectively the transverse steel content, the longitudinal steel content and the steel profiles contribution ratio – see the definition in Section 4.10.1; ω_{wd} is the mechanical volumetric ratio of confining hoops within the boundary zones as defined in Section 4.10.2; θ_y and θ_u are respectively the yield and ultimate drift ratio calculated as $\theta_y = \Delta_y/H$ and $\theta_u = \Delta_u/H$, where H_w denotes the height of the central axis of the horizontal actuator relative to the wall base; Δ_y and Δ_u are respectively the displacements measured at the load application level at yield and ultimate displacement; the ductility ratio μ_Δ is calculated as $\mu_\Delta = \Delta_u/\Delta_y$.

Because the values of parameters like Q_y and θ_y are necessarily conventional, because there are small differences in the specimens and because the

Table 4.1 Characteristics of wall specimens and test results

Reference, and number and type of test	λ	ν_{exp}	ρ_t %	ρ_l %	δ_a	ω_{wd}	θ_y %	θ_u %	μ_Δ	α_{Eclg}
(Cho et al., 2004) Two cyclic	3.7	0.10	0.46 0.57	0.52	0.35	0.0	0.67	2.7	4	0.35 0.43
(Zhou et al., 2010) Three cyclic	3.7	0.09 0.18	0.57	0.75	0.37	0.084	0.35	2.5	7	0.33
(Zhou et al., 2010) Nine cyclic	1.5 2.0	0.09 0.18	0.57	0.75	0.37	0.084	0.45	1.7	4	0.22
(Zhou et al., 2010) Three cyclic	0.8	0.18 0.24	0.57	0.75 1.10	0.37	0.084	0.47	1.0	2	0.05
(Qian et al., 2012) Six cyclic	2.4	0.55 0.73	0.42	0.79	0.15 0.22	> 0.2	0.30	1.4	4	0.45
(Dan et al., 2011a) Six cyclic	2.6	0.15 0.21	0.37	1.60	0.20 0.26	0.084	0.86	4.3	5	0.16
(Ji et al., 2015) Five cyclic	2.3	0.60	0.33 0.42	0.7	0.27	0.24 0.29	0.30	1.8	6	0.68
(INSA, 2017) Six monotonic	2.1	0.00	0.62 1.23	1.12	0.31	0.31* 0.62*	0.80	4.8	6	0.24
(ULiege, 2017) Six monotonic	2.1	0.00 0.13	0.66	1.19	0.31	0.30	0.74	4.5	6	0.30 0.45
(Wu et al., 2018) Six cyclic	2.4	0.19 0.13	0.70	0.52	0.14	0.0	0.60	3.0	4	0.18
(Chrzanowski, 2019) Two monot.	5.2 4.1	0.00 0.00	0.59	0.38	0.33	0.11	1.30	5.2	4	0.25 0.17

(Continued)

Table 4.1 (Continued) Characteristics of wall specimens and test results

Reference, and number and type of test	λ	ν_{exp}	ρ_t %	ρ_l %	δ_a	ω_{wd}	θ_y %	θ_u %	μ_Δ	α_{Edg}
(Bogdan et al., 2019) Four cyclic **	**	0.65 0.70	0.13	0.54	0.48	0.38	0.72	1.4	2	> 1
(Zhang et al., 2020) Four cyclic	2.0	0.4 0.6	0.67	0.67	0.36	0.17	0.80	3.5	4	0.25
(Zhou et al., 2021) Six cyclic	2.4	0.35	1.34	1.04	0.22 0.44	0.18	0.43	2.2	> 5	0.61
(Zhou et al., 2021) Four cyclic	2.4	0.23	1.12	0.93	0.44	0.18	0.43	2.2	> 4	0.49

* hoops in boundary zones are not closed hoops

** column tests with high λ for which the applied shear V is close to 0 because M results from a progressively increased axial force N of constant eccentricity so that ν_{exp} grows with N; the value of ν_{exp} given in Table 4.1 correspond to maximum N applied in the tests.

intention is to present a short synthesis, the values in Table 4.1 are an order of magnitude for each test series. The reference publications should be consulted for more detailed information.

4.9.2.1 Observations on the coefficient α_{EcIg} deduced from tests on hybrid walls

Table 4.1 shows that:

- α_{EcIg} is greater than 0.5 only for the tests in which the axial force ratio $\nu_{d,c}$ is greater than 0.45 and the confinement parameter ω_{wd} greater than 0.2 (Qian et al., 2012; Ji et al., 2015).
- α_{EcIg} is greater than 1.0 only for the tests in which the axial force ratio $\nu_{d,c}$ is greater than 0.65, the confinement parameter ω_{wd} greater than 0.2 and the steel profiles contribution ratio δ_a greater than 0.45 (Bogdan et al., 2019); α_{EcIg} greater than 1.0 is possible only because $E_c I_g$, which represents only concrete, is not a representative parameter of the stiffness of sections in which there is a large contribution of steel profiles.
- α_{EcIg} is around 0.2 for several specimens, thus below the lower bound 0.35 of AISC341-16 or ACI 318-19 and also below 0.5 suggested in Eurocode 8 (2022); the deformation due to shear which is implicitly present in α_{EcIg} contributes to low values of that parameter (Zhou et al., 2010; Dan et al., 2011a; Wu et al., 2018; Zhang et al., 2020).
- The lower α_{EcIg} seem linked to either a low confinement parameter ω_{wd} (Wu et al., 2018).

or to tests in which *D regions* predominate, typically beam tests simulating walls (INSA, 2017; Chrzanowski, 2019) or tests on the cantilever in which the Δ_{slip} has not been brought to a minimum by adequate constructional measures (anchorage of profiles at free end, spreader beam).

- α_{EcIg} is very low for walls with an aspect ratio under 1.5 (Zhou et al., 2010).
- The scatter in the values of α_{EcIg} may be in part due to differences in flexibility of the connections resisting longitudinal shear at profiles-concrete interfaces; that connection is stiff if the resistance by bond combined with friction is greater than the longitudinal shear observed when plastic bending takes place, this is more easily the case with encased steel tubes (Qian et al., 2012) because the resistance τ_{Rd} to longitudinal shear is greater for tubes than for open sections.
- The values of the parameter α_{EcIg} in Table 4.1 are mostly derived from tests on walls of relatively low slenderness ($\lambda < 4$), in which the deformed shape is much influenced by the rotation in the plastic hinge

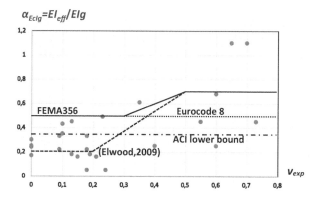

Figure 4.35 Experimental values of α_{Eclg} and code definitions for hybrid walls and columns

zone at the bottom of the walls; in those tests, the part of a wall which remains in the elastic range is of the same order as the plastic hinge zone; in a building, a much greater part of a wall remains in the elastic range and its global stiffness is greater than the one deduced from tests on relatively short specimens. For this reason, it can be considered that the range of α_{Eclg} defined in codes for reinforced concrete components, which are greater α_{Eclg} than those given by the direct observation of experimental results, may be kept for hybrid components.

4.9.2.2 Bending stiffness of components for design

Once the reinforcement by bars and by encased profiles are defined, better estimates of the stiffness of components better than $EI_{eff} = \alpha_{Eclg} E_c I_g$ can be calculated and a more refined analysis of the sections and of the structure performed. Equation (3.2) for $(EI)_{eff,II}$ of Eurocode 4 (2004) provides fair estimates of the real stiffness of components taking into account implicitly concrete cracking, shear deformations and bar slip at anchorage; it may be used to determine the internal forces in a structure by means of a first-order analysis.

4.9.3 Moment of resistance, transverse shear and ductility

4.9.3.1 General conclusions from testing activity concerning the moment of resistance

The test results summarized in Table 4.1 all correspond to specimens of walls designed to yield in bending prior to shear cracks. To reach this objective, all specimens have overstrength in shear resistance, at different degrees, with one exception: the squat walls with λ=1.5 in Zhou et al. (2010) which, due to their low shear span ratio, fail essentially in shear. As can be seen in Table 4.1, in which most test results have been obtained under cyclic loading, the ductility factor associated with flexural yielding is at least 4, even if the axial compression ratio ν_{exp} is high. The specimens also generally demonstrate a reserve of strength above the calculated yield moment. This confirms the significance of the "weak bending resistance – strong shear resistance" design concept in slender walls. Some other specific features of the various test programs are summarized hereafter.

4.9.3.2 Tests on hybrid walls in Ji et al. (2010) and in Zhou et al. (2010)

The results of cyclic tests in Ji et al. (2010), Qian et al. (2012) and in Zhou et al. (2010) are interesting because ductility in bending is developed without connectors on the encased steel profiles or concrete-filled steel tubes embedded in the walls; this allows calibration of the design method proposed in Sections 4.5 and 4.7.

The six hybrid wall specimens in Ji et al. (2010) or Qian et al. (2012) have been submitted to cyclic tests, and their displacement ductility ratio μ_Δ are all greater than 4.6 though the axial load ratio ν_{exp} is above 0.56. Boundary elements confined by stirrups such that ω_{wd} was greater than 0.2 contributed to that result.

The specimens with aspects ratio greater than 1.5 described in Zhou et al. (2010) and in Ji et al. (2010) or Qian et al. (2012) realize the ductility conditions given in Section 4.7.4: $V_{Rd,max}$ is 1.8 to 2.8 times greater than $V_{Rd,s}$ and 1.75 to 1.10 times greater than Q_y. It can be pointed out that, in many of these tests, $V_{Rd,s}$ represents only 40 to 50 % of Q_y, so the walls would have failed in shear without the contribution $V_{Rd,a}$ of the encased steel sections to shear resistance. As $V_{Rd,s} + V_{Rd,a} \approx Q_y$, a plastic mechanism combining bending and shear is formed. It can also be shown, by means of

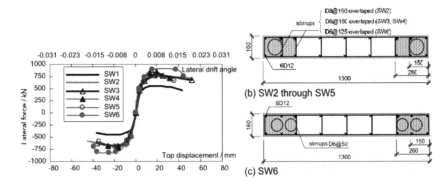

Figure 4.36 Envelope of experimental curves and sections of five tested specimens (Ji et al., 2010)

a comparison of the overall resistance of hybrid walls to the one of a reference wall reinforced only by bars, that the contribution of the embedded steel profiles to shear resistance is equal to their calculated plastic shear resistance. The contribution of the embedded steel profiles is effective up to a displacement corresponding to the maximum resistance and over, up to a failure by concrete crushing at the base of the walls.

In some tests, the effective shear resistance at the yield load in bending is 25% greater than the calculated resistance to shear. This overstrength may be explained by the strain hardening of the encased profiles submitted to cyclic plastic shear, to which corresponds a well-known increase in the apparent shear yield stress from $f_y/\sqrt{3}$ up to f_y. The experimental and calculated yield moments compare well for the specimens with aspect ratios $H/l_w \geq 1.5$.

The comparison of bending moments is not satisfactory for specimens with an aspect ratio $H/l_w \leq 0.8$ because they fail in shear, not in bending. Beam or truss models are inadequate for squat walls because these behave like corbels and should be assessed as such, with one active vertical steel section and one compression strut. The ratio of resistance in experiments to resistance calculated by a classical struts and ties model for the specimens in Zhou et al. (2010) is about 1.25.

4.9.3.3 Tests on hybrid walls in Dan et al. (2011a)

There are shear connectors on all steel profiles. The applied shear Q_y at yield in bending is on average 40% of the calculated resistance V_{Rd} of the walls to transverse shear. Due to cyclic loading, alternate diagonal cracks appear after the walls yield in bending. The maximum applied load V_{max} is

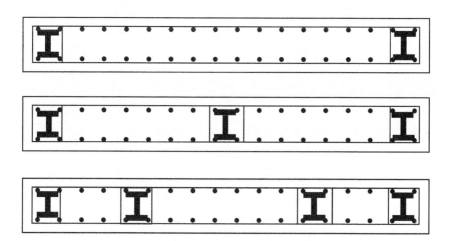

Figure 4.37 Types of walls in (Zhou et al., 2010)

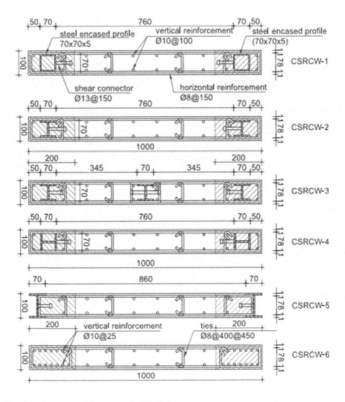

Figure 4.38 Specimens in (Dan et al., 2011a)

equal to 1.4 to 1.6 times the yield load Q_y and, as in Ji et al. (2010), V_{max} is obtained for displacements more than four times greater than the yield displacement. The "weak bending resistance – strong shear resistance" design provides this ductility and reserve of strength.

4.9.3.4 Tests on hybrid walls in Ji et al. (2015)

The sections of the walls tested by Ji et al. (2015) are shown in Figure 4.39. The calculated shear resistance of the walls is greater than the applied shear, so that plastic bending, with cracks perpendicular to the wall axis, is the first ultimate limit state reached and that ductility is present. As the applied shear at yield Q_y is a significant portion of the shear resistance V_{Rd} and because shear at the maximum load V_{max} is almost equal to the shear resistance V_{Rd}, diagonal cracks appear after the elements yield. This correspondence between observations and calculation of the shear resistance confirms the validity of the design expressions given in Sections 4.5 and 4.7. The maximum applied loads V_{max} are equal to 1.25 times Q_y.

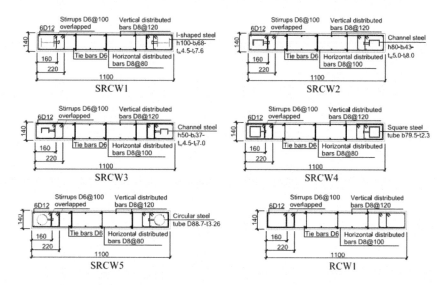

Figure 4.39 Test specimens in (Ji et al., 2015)

4.9.3.5 Tests on hybrid walls at INSA Rennes in Degée et al. (2017)

The tests at INSA Rennes presented in Degée et al. (2017) are three points bending tests on beams. They are designed to set forward the influence of constructional details like the type of shear connectors, headed studs or welded plates, and the density of stirrups, classified normal or high with in that case a notation HC in the specimen's name. There are no closed hoops in the boundary zones (see Figure 4.40). The flanges of steel profiles are parallel to the sides of the wall. An extra length of beam is provided to anchor the steel profiles in the support zones.

The load-displacement curves show that the same stiffness and yield load are realized with headed studs or welded plates shear connectors; the behaviour is slightly more ductile with the greater density of transverse reinforcement (see Figure 4.41, noting that the applied shear V is equal to 0,5 times the applied load). Without shear connectors (specimens BW and BWHC), the yield strength is about 20% smaller than with; no relative slip between concrete and profile is observed until a failure of brittle nature takes place, leading to a ductility factor of only 2.

The calculations indicate that, for all specimens, $V_{Rd,s}$ is greater than $V_{Rd,max}$. The condition $V_{Rd,s} < V_{Rd,max}$, which is necessary to achieve ductility if the moment of resistance in bending is reached simultaneously with the resistance in shear, is not realized. However $(V_{pl,Rd,tot} + V_{Rd,s})$ or $(V_{pl,Rd,tot} + V_{Rd,max})$ are more than three times greater than Q_y so that a ductile

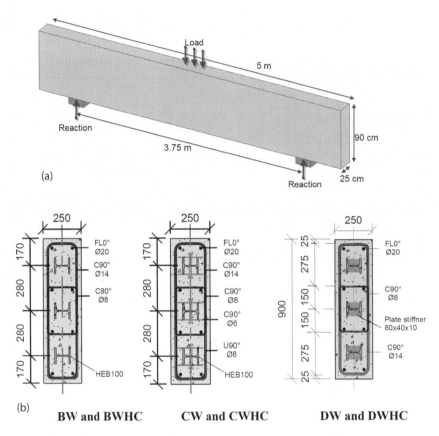

Figure 4.40 Top: test configuration. Bottom: cross-sections of specimens (Degée et al., 2017)

bending is realized first and shear cracks appear only for greater applied displacements.

It is interesting to note that, in agreement with the calculation method proposed in Section 4.7, it is not necessary to require that $V_{Rd,max}$ be greater than Q_y to reach the design strength in bending and to achieve ductility. This is due to the contribution $V_{pl,Rd,tot}$ of the encased steel profiles to the resistance to transverse shear.

The measured shear stresses outside of the disturbance zones D, which means at mid-distance between the load application point and support, correspond well with the calculations. The same comment applies to the distribution of shear between the reinforced concrete truss and the steel profiles.

Shear strains measured in the steel profiles are similar in all specimens but twice greater in the lower steel profile than in the mid-height and top profiles on the same vertical line. Above the supports, shear stresses are

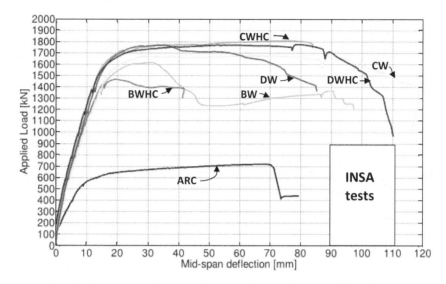

Figure 4.41 Load-deflection curves of the specimens described at Figure 4.40 (Degée et al., 2017)

very low: in the absence of a stiffening element spreading the reaction force in the complete section depth, the upper corner of the beam is a "dead triangle" in the stress field. This results from the localized character of side forces at the point of application of loading and at supports. It may be considered that the specimens are essentially constituted of two "D type" regions.

4.9.3.6 Tests on hybrid walls at ULiege in Degée et al. (2017)

The specimens tested at ULiege presented in Degée et al. (2017) are cantilever walls anchored in a reinforced concrete foundation block and submitted to a shear force V applied through a reinforced concrete beam which stiffens the wall, simulating the standard layout at storey level in a building. A constant 1000 kN axial load N is applied to specimens CSN and DSN. In the hybrid specimens, three HEB100 steel profiles are encased, with webs parallel to the wall faces. One reference reinforced concrete wall with the same reinforcement by bars as in the five hybrid specimens is also tested – see Figure 4.42. The specimens are designed to set forward the influence of constructional details, out of which some do not comply with seismic requirements like the confinement of the boundary zones of walls. The diagrams of horizontal force V versus displacement recorded in the tests are given in Figure 4.42.

Only the reinforced concrete specimen ARC and the hybrid specimens CS and CSN with headed studs connectors behave in a ductile way, where

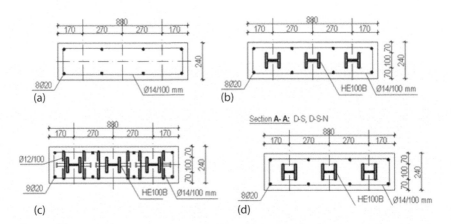

Figure 4.42 Respectively from top to bottom the cross-sections of specimens ARC, BS, CS and CSN, DS and DSN (Degée et al., 2017)

the moment of resistance of CSN is greater than the one of CS due to the application of an axial force N, while the ductility of CSN is smaller. The maximum applied force corresponds either to plastic bending, with rebars and profile yielding in tension (specimens CS and CSN) or to a failure mode related to poor resistance to longitudinal shear at steel profile-concrete interface (specimens BS, DS and DSN).

The conclusions on the relative values of $V_{Rd,s}$, $V_{Rd,max}$, $(V_{pl,Rd,tot} + V_{Rd,s})$, $(V_{pl,Rd,tot} + V_{Rd,max})$ and Q_y are the same as for the tests at INSA detailed above. As a consequence, the shear stress τ_a measured in the steel profiles is low enough to have no influence on their axial capacity. Specimens DS and DSN with plate connectors suffered early failure because there were no transverse stirrups around the embedded profiles to resist the outward thrust of the local compression struts at plate connectors when H profiles are oriented, as shown in Figure 4.42. The correspondence of measured to calculated shear stresses is excellent for specimens CS and CSN in which the necessary resistance to longitudinal shear is provided by shear studs: the average ratio of measured to calculated shear in the profiles is equal to 0.94.

4.9.3.7 Tests on hybrid walls in Wu et al. (2018)

The experimental activity of Wu et al. (2018) focused on developing precast hybrid walls. One reference reinforced concrete wall and six hybrid walls are tested. The encased profiles are CFST on which the transverse reinforcement is welded, and shear connectors taking the form of an arch are also welded on the tubes. There is no confinement in the boundary zones – see Figure 4.44.

The horizontal and vertical loads are applied through a reinforced concrete beam. The wall is connected at its basis to a strong reinforced concrete beam. The connection consists of steel tube sleeves welded to horizontal H

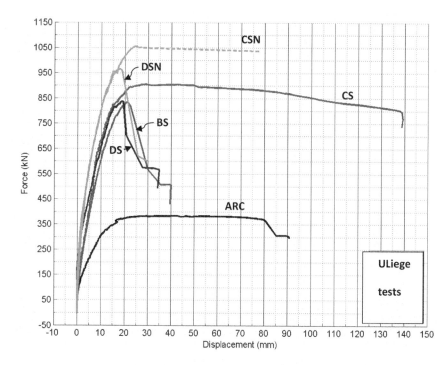

Figure 4.43 Diagram of applied force *V* versus horizontal displacement of the specimens shown at Figure 4.42 (Degée et al., 2017)

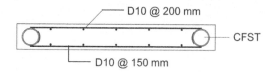

Figure 4.44 Test specimens in (Wu et al., 2018)

sections embedded in the basis, and the encased CFST are inserted in the sleeves. Both the sleeves and the tubes have surface indentations realized by stack welds and the tension/compression force is transmitted through an injected grout. All specimens were designed for flexure-dominant behaviour and behaved in a ductile way. The wall anchorage system was proved valid. However, due to the lack of longitudinal and transverse reinforcement in the boundary zones, premature visible horizontal cracking was observed at each horizontal reinforcement level for a relatively low 0.5% drift.

4.9.3.8 Tests on hybrid walls in Zhang et al. (2020)

The four hybrid walls in Zhang et al. (2020) are made of high-strength materials: C60 concrete reinforced by steel fibres together with S600 or

S1200 SRUSH longitudinal bars; encased S460 steel tubes with fibre-rein-forced concrete infills; no shear connectors are present. The tests demon-strate that high-strength walls can provide the calculated bending moment together with ductility. Because the steel tubes represent 80% of the ten-sioned steel in the boundary zones, the yield moment is relatively indepen-dent of the other longitudinal steel reinforcement characteristics. Flexural cracks appear first, which extend obliquely as the drift ratio increases. The residual deformation after unloading is small up to a loading drift ratio of 2%; it is smaller with S1200 bars in boundary elements than with S600 bars. Adding steel fibres to the concrete matrix improves the deformation capacity and allows full use of the performance of high-strength steel bars: without fibres, the ductility of walls with high-strength steel bars is poor. The measurements show that the plane section practically remains plane.

4.9.3.9 Tests on hybrid walls in Zhou et al. (2021)

The external dimension of the ten hybrid walls in Zhou et al. (2021) is 900x150 mm. The external encased sections are CFST of 102x4 mm dimensions. In one wall specimen, only external profiles are present; in the other nine specimens, there are one or two internal CFSTs and different layouts in terms of CFST diameters and shear reinforcement. The walls are made of relatively high-strength concrete, C50 concrete for the main section and C80 for the infills. Shear connectors are ring bars of diameter 6 welded on the CFST at a 300 mm distance. The shear resistance brought by reinforced concrete is much greater than the shear action effect at a yield of the specimens in bending: $V_{Rd,s}$ and $V_{Rd,max}$ are respectively 1.6 and 2.3 times greater than $V_{E,exp}$ and 2.75 and 4 times greater than $V_{E,RC}$ so that no failure in shear is observed; the specimens demonstrate a ductility in bend-ing in the range of 4.3 to 6.8. The shear connections work well up to the yield load, but at twice the yield displacement, diagonal cracking is visible around the internal CFST; this fits with the observation that longitudinal shear action effect is about twice greater at internal encased sections than at external ones – see Sections 4.5.6 and 4.7.2, but it may also be due in part to a relatively small concrete cover in some specimens. But for that aspect, the observations are positive: the specimens are stiff up to yield load, with $\alpha E_c I_g = 0.49 E_c I_g$ for $\nu_{d,c} = 0.23$ and $\alpha E_c I_g = 0.61 E_c I_g$ for $\nu_{d,c} = 0.35$; the shear ring bars are effective for the external CFST; all specimens have a stable hysteresis performance. A general conclusion is given: with high-strength encased concrete CFST, the allowable axial load ratio could be increased above the standard values in application, presumably up to 0.70.

4.9.3.10 Tests on hybrid columns in Bogdan et al. (2019)

The tests in Bogdan et al. (2019) refer to hybrid columns in which the encased profiles are hot rolled jumbo sections HEM100 (120x106x12x20 mm) – see Figure 4.27; they assess the great load-bearing capacity of real hybrid columns well. Static compression loads with a constant eccentricity are applied which submit the specimens to a combination of axial force and bending without shear. The test results shown in Figure 4.27 show the validity of the different possible methods in calculating the resistance of hybrid columns submitted to a combination of axial force and bending.

4.9.4 Longitudinal shear at steel profiles interface with concrete

Concerning longitudinal shear at the concrete-steel profiles interface, the results of cyclic tests on six wall specimens in Qian et al. (2012), on 13 specimens in Zhou et al. (2010) and on four wall specimens in Zhang et al. (2020), all with aspects ratio greater than 1.5, are interesting to test the validity of the design expressions in Sections 4.5 and 4.7, because ductility in bending is developed without connectors on the encased steel profiles. The use of the design expressions in Section 4.7.2 indeed indicates that, for the specimens in Qian et al. (2012) and Zhang et al. (2020), the calculated design resistance to longitudinal shear provided by bond plus friction is greater than the applied longitudinal shear corresponding to the external force Q_y at yield of the specimens: $V_{Rd,total} > V_{l,a}$. For the specimens in Zhou et al. (2010), $V_{Rd,total}$ is, at most, 20% smaller than $V_{l,a}$; however, the absence of shear connectors in the test specimens did not cause trouble, probably because the real resistance to longitudinal shear is greater than the design one by a factor of the order of 1.5 so that $V_{R,total,real}$ is about $1.3V_{l,a}$. It can be observed that safety is better realized with encased CFST than with encased open sections like H because the resistance τ_{Rd} to longitudinal shear is significantly greater for CFST than for open sections: 0.55 N/mm^2 instead of 0.30 N/mm^2 according to Eurocode 4 (2004) (see Table 2.2) can be observed also that, if friction did not exist, the specimens would have failed in longitudinal shear much before ductility in bending was realized. The practical conclusion is that if friction was not considered in design, shear connectors would appear necessary in all specimens. Problems would exist if the bond resistance was smaller than 0.3 N/mm^2 and the friction coefficient smaller than 0.5, which would happen if there was loose rust or paint on the profile surface. A line in Table 2.2 recalls this.

The safe-sided character of μ and τ_{Rd} might even be greater in Zhang et al. (2020) than mentioned above because the code values taken into account in the calculations are meant for concrete without fibre reinforcement: fibres, which prevent concrete cracking, probably increase μ and τ_{Rd}.

For the specimens in Dan et al. (2011a), the calculations show that for all specimens except the one with partially encased H sections, the total resistance $V_{Rd,l,a}$ to longitudinal shear at concrete-steel profiles interface provided by bond and friction is greater than the calculated action effects $V_{Ed,l,a}$ at yield load level; shear connectors might have been not necessary.

For the specimens in Ji et al. (2015), the calculated resistance to longitudinal shear provided by bond alone or by friction alone is smaller than the action effects, but the total V_{Rd} provided by bond and friction is greater than the action effects $V_{l,a}$ at yield load level; shear connectors might have been unnecessary, in particular with encased CFST.

For the specimens in INSA and those at ULiege in SMARTCOCO (2017), the calculations indicate that the total resistance $V_{Rd,l,a}$ to longitudinal shear at the concrete-profile interface provided by the combination of bond and friction is smaller than the applied longitudinal shear force $V_{Ed,l,a}$ corresponding to the theoretical yield load in bending. This need for shear connectors predicted by the calculations is confirmed by the brittle and premature failure of walls specimen BS in which no connectors were present. The calculations also correctly indicate that the transverse steel content which is greater in specimen BWHC than in specimen BW contributes to a stiffer reinforced concrete truss and to a greater transverse shear $V_{Ed,RC}$ taken by the reinforced concrete truss; as the longitudinal shear is increased accordingly, the ratio of resistance to action effects for longitudinal shear is less favourable for specimen BWHC than for specimen BW, which illustrates less favourable load-displacement curves for specimen BWHC – see Figure 4.41.

4.10 STEEL CONTENT IN HYBRID COLUMNS OR WALLS

4.10.1 Longitudinal steel

In design codes, there are limitations to the content of longitudinal steel in reinforced concrete columns or walls and in composite steel-concrete columns. Research has not focused on the applicability of these limitations to hybrid walls or columns, but valuable information is available because the test specimens have generally been designed outside of consideration to code limitations.

The parameters for longitudinal steel in hybrid columns or walls are:

- The percentage of longitudinal reinforcement by bars: $\rho_s = A_s / A_c$, where A_c is the gross area of concrete.
- The percentage of longitudinal reinforcement by profiles: $\rho_a = A_a / A_c$.
- The total percentage of longitudinal steel: $\rho_{s,tot} = \rho_s + \rho_a$.
- The steel profiles contribution ratio: $\delta_a = A_a f_{yd} / N_{pl,Rd}$ where $N_{pl,Rd}$ is the plastic resistance to compression of the composite column or wall calculated with Equation (4.34).

Limitations on the values of these parameters have different objectives:

- The lower bounds intend to avoid the fragile behaviour of concrete.

- The upper bounds intend to avoid geometrical congestion of reinforcement.

For composite steel-concrete columns, Eurocode 4 (2004) prescribes: $\rho_s \geq 0.3\%$ and $0.20 \leq \delta_a \leq 0.90$

For reinforced concrete components, Eurocode 2 (2004) prescribes:

- $\rho_s \geq 0.2\%$ or $\rho_s \geq 0.10\ N_{Ed}/f_{yd}$, whichever is the greatest, where f_{yd} is the design yield strength of the reinforcing bars and N_{Ed} is the design axial compression force.
- $\rho_s \leq 4\%$ outside lap locations and $\rho_s \leq 8\%$ in lap locations.

The range of the parameters ρ_s, ρ_a, δ_a and $\rho_{s,tot}$ realized in the over 50 tests of the experimental programs summarized in Section 4.9 and the compliance of these parameters to code limitations may be summarized as follows:

- ρ_s range from 0.7% to 2.5% and comply with Eurocode 2 (2004) and Eurocode 4 (2004) limitations.
- ρ_a range from 0.9% to 13.8% and does not comply with the upper bound of ρ_s of Eurocode 2 (2004).
- δ_a range from 0.04 to 0.70 and does not comply with the lower bound 0.20 prescribed in Eurocode 4 (2004).
- $\rho_{s,tot}$ range from 7.8% to 18.4% and does not comply with the upper bound of ρ_s of Eurocode 2 (2004).

Thus, tests show that hybrid columns or walls in which the steel content does not comply with code requirements of Eurocode 2 (2004) and Eurocode 4 (2004), which are not directly conceived for them, provide the calculated design resistance and behave in a ductile way. There are explanations.

Neither $\rho_a = 13.8\%$ nor $\rho_{s,tot} = 18.4\%$ complies with the upper bound of $\rho_s = 4\%$ of Eurocode 2 (2004). They should not, because the upper bound to the longitudinal steel content is meant to avoid congestion of reinforcement by bars and because encased steel profiles are used as reinforcement to allow a greater steel content without congestion.

The observed steel profiles contribution ratio $\delta = 0.04$ is smaller than 0.20 and thus it does not comply with the lower bound limitation of Eurocode 4 (2004). The objective of this limitation in Eurocode 4 (2004) is to avoid design rules for composite columns being applied to components with such a small single central steel profile, where they behave like a reinforced concrete column and are out of the scope of Eurocode 4 (2004). However, in hybrid components, there are several steel profiles intended to resist bending moments in a way similar to bars. The objective may be more compact components with greater performance; it may also be a designer's choice to use small H-, L- or U-encased sections rather than bars to solve the problem of availability of bars on the market. In both cases, the lower bound $\delta = 0.04$ of composite columns is out of context and need not be applied.

Based on those considerations, it is suggested that longitudinal steel in hybrid components should comply with a), b) and c):

a) Encased steel profiles in hybrid walls or columns should satisfy: $\delta_a \leq 0.90$:
b) The minimum and maximum ratio ρ_s of longitudinal reinforcement by bars, the minimum number of longitudinal bars, the clear spacing between longitudinal bars and the concrete cover for longitudinal reinforcement should comply with the rules in the reference reinforced concrete code for the intended design action. In particular for ρ_s, in the frame of Eurocode 2 (2004): $0.2\% \leq \rho_s \leq 4\%$:
c) The clear spacing between longitudinal bars and encased steel sections and the concrete cover should comply with the reference code for composite structures for the intended design action.

4.10.2 Transverse steel

4.10.2.1 General

Transverse reinforcement should be designed to provide resistance against shear action effects in concrete and in encased profiles calculated in the way explained in Sections 4.4 or 4.5. The resistance against these shear action effects should be established as explained in Sections 4.6 and 4.7.

The rules for transverse reinforcement by bars prescribed in the reference reinforced concrete design codes for the considered design action (static, earthquake) should be applied to hybrid components. Transverse reinforcement should be continuous from one edge of the wall to the opposite edge and anchored in the boundary zones (Figure 4.45). These general indications deserve the specific comments given hereunder.

4.10.2.2 Confinement of boundary zones of walls

Encased steel open sections like H or I realize local confinement of concrete between the flanges for what concerns local compression struts around shear connectors (see Section 2.3), but there is no such confining effect for what concerns the longitudinal strains due to the compression and bending applied to the wall. Indeed, in the elastic range, Poisson's coefficient of steel (≈ 0.3) is greater than the one of concrete (≈ 0.2), so a general state of compression in the hybrid section induces more lateral expansion of the profiles than of the surrounding concrete. This generates a circular tension force in the concrete around the profile which would generate radial cracking in the absence of specific reinforcement. Closed stirrups around the profiles resist that tension and create a radial compression on the profiles that fosters resistance to longitudinal shear by friction and bond at the interfaces between steel profiles and concrete. Once concrete and steel enter the plastic domain, Poisson's coefficients of both materials become greater than 0.3, but the effect of confinement remains necessary.

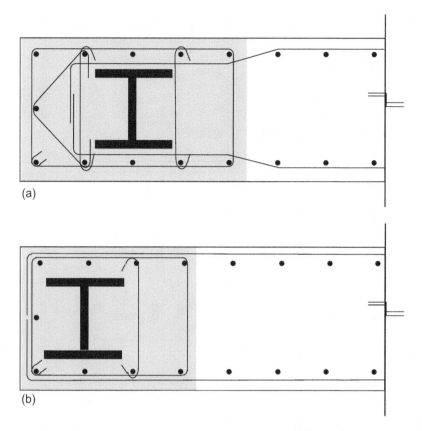

(a)

(b)

Figure 4.45 Layout of transverse reinforcement. Top: for high ductility. Bottom: for basic
ductility. The shaded area is the boundary zone

Another mechanical effect requires confining reinforcement or ties transverse to the wall: as explained in Section 4.5.6 an external encased profile
attracts an important part of the compression strut force F_{comp} so that a
deviation of the compression force takes place that involves a tension force
perpendicular to the wall plane, as shown in Figures 4.46a and b.

With a H section with a supporting dimension equal to half of the wall
thickness, this transverse tension force F_{tie} over the height $z\cot\theta$ of the "reference cell" may be assessed with Equation (4.53) where all symbols are
defined as in Section 4.5.6:

$$F_{tie} = 0.125 \times \left(F_{comp} \times A_{prof} \right) / \left(A_{prof} + A_{bars} \right) \tag{4.53}$$

Confining hoops and ties should resist those tension forces within a wall
length approximately equal to its thickness. Besides the specific indication for hybrid walls, the requirements for confining reinforcement, hoops,

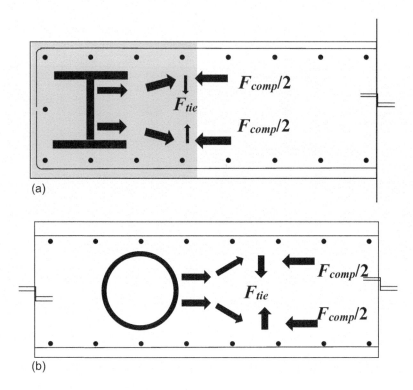

Figure 4.46 Deviation forces in the vicinity of an encased H section (top) and an encased circular section (bottom)

stirrups and cross ties should not be smaller in hybrid sections than in reinforced concrete or composite sections: it is advised to place around the external steel profiles no less than the confining reinforcement imposed around composite steel-concrete columns for static design (Eurocode 4 (2004), AISC360-16) or for earthquake-resistant design in (Eurocode 8 (2004); AISC341-16). In the latter case, they are vital to achieving ductility in the boundary regions of the walls. Preventing the buckling of longitudinal reinforcement may require diagonal ties or hoops, as can be seen in Figures 4.1 and 4.45a. And if shear connectors are used, stirrups or links should be placed at each connector and they should be designed to resist a tension force equal to the shear capacity of the connector.

In Eurocode 8 (2022), the confinement of the boundary regions is characterized by the mechanical volumetric ratio of confining hoops within the critical regions ω_{wd} calculated with Equation (4.54) :

$$\omega_{wd} = \frac{volume\,of\,confining\,hoops}{volume\,of\,concrete\,core} \times \frac{f_{yd}}{f_{cd}} \tag{4.54}$$

ω_{wd} should satisfy a condition related to the demand of ductility, with ω_{wd} not smaller than 0.05 for medium ductility and not smaller than 0.08 for high ductility.

4.10.2.3 Confinement around internal encased sections

The internal encased sections are active parts of the concrete compression struts which participate in the wall's or column's resistance to transverse shear – see Section 4.5.3. Being "hard spots", they influence the field of stresses in concrete. Links placed on both sides of encased section are needed to resist the local deviation of the compression struts forces suggested in Figure 4.46 a). Encased H sections offer a support that is perpendicular to compression struts forces and, in the absence of studies on the subject, it may be proposed that the transverse tension force F_{tie} over the height $z\cot\theta$ of the "reference cell" be assessed with Equation (4.55):

$$F_{tie} = 0.125 \times F_{comp} / \eta \tag{4.55}$$

where η is defined by Equation (4.21), (4.22) or (4.23).

Encased circular sections offer cylindrical support to compression forces which is effective on a fraction of the diameter, so that, as qualitatively shown in Figure 4.46 b, the deviated forces are greater in that case than with H sections. Therefore, it is suggested that transverse ties in the vicinity of circular tubes be designed to resist F_{tie} calculated with Equation (4.56):

$$F_{tie} = 0.2 \times F_{comp} / \eta \tag{4.56}$$

4.10.2.4 Concrete cover

The cover of concrete around the encased steel sections should be such that stirrups constitute effective confinement and ensure the safe transmission of bond forces, the protection of steel profiles and bars against corrosion and prevent concrete spalling. This is even more important with encased profiles due to the uncertainty on the exact field of stresses around them explained above. Insufficient cover may lead to dense local cracking and to local failure at the ultimate stage (Zhou et al., 2021). The concrete cover should satisfy the rules of composite steel-concrete structures and of reinforced concrete structures.

4.11 DESIGN EXAMPLES FOR HYBRID WALLS

4.11.1 Moment M-axial force N interaction curve by the simplified method

4.11.1.1 Geometrical data and strength properties of the wall

Wall section: l_w=1600 mm b_w=400 mm Gross section $l_w\ b_w$ = 640000 mm^2

Steel profiles: HE240B steel grade S460 A_a=10600 mm^2 $f_{yd,a}$= 460/γ_{M0}=460 N/mm^2

$b = h$ = 240 mm t_f = 17 mm t_w= 10 mm

Distance from external profile axis to axis of symmetry of the hybrid section: d_s = 600 mm

Vertical reinforcement

- top and bottom of wall section: A_{s1} = 3 diameter 20 mm = 3 x 314 = 942 mm^2
- each side of wall section: A_{s2} = 9 diameter 12 mm = 9 x 113 = 1018 mm^2
- $f_{yd,s}$= 500/γ_s= 500/1.15= 435 N/mm^2

Concrete grade C50 $f_{cd} = f_{ck}/\gamma_c$ = 50/1.5 = 33.3 N/mm^2

$0.85 f_{cd}$ = 28.3 N/mm^2 E_{cm}=37000 MPa

Concrete area $A_c=l_w$ x b_w − 3 A_a − 2A_{s1} − 2A_{s2}=640000−3x10600−2x942−2x1018= 604280 mm^2

4.11.1.2 Equivalent plates

It is explained in Section 4.6.3 how a line of n rebars can be replaced by an equivalent plate in order to simplify the calculations.

Plate N°1 equivalent to the top and bottom layers of 3 diameter 20 bars

- Distance between most distant bars d_b = 300 mm.
- Equivalent plate length L_{p1} = 300 x 4/3 = 400 mm.
- Equivalent plate thickness t_{p1} = A_{s1} /400 = 942/400 = 2355 mm.
- Distance of plate centre to axis of symmetry: d_{p1} = 5 x 150 = 750 mm.

Plate N°2 equivalent to one side layer of 9 diameter 12 bars

- Distance between most distant bars d_b = 1200 mm.
- Equivalent plate length L_{p2} = 1200 x 10/9 = 1333 mm.
- Equivalent plate thickness t_{p2} = A_{s2} /400 = 1018/1333 = 0.764 mm.
- Distance of plate centre to axis of symmetry: d_{p2} = 0 mm.

Rectangle equivalent to one HE240B: height H_{eq}= b = 240 mm $B=A_a/H_{eq}$=10600/240=44.17mm

4.11.1.3 Squash load – pure compression – key Point A

$N_A = A_{a,total} \times f_{yd,a} +0.85 f_{cd} A_c + A_{s,total} \times f_{yd,s}$

- = 3 x 10600 x 460 + 0.85 x 28.3 x 604280 +2 x (1018 + 942) x 435
- = 14628000 + 14535955 + 1705200 = 30869155 N = 30869 kN

4.11.1.4 Plastic neutral axis and moment of resistance in pure bending – key Point B

By definition, $N_B = 0$.

The PNA in pure bending is tentatively placed between the upper and mid-section HE240B, at an unknown distance hn from the axis of symmetry of the section (see Figure 4.47). hn is determined by expressing that the resultant compression resistance (above PNA), and the resultant tension resistance (below PNA) is equal in the case of pure bending.

Plastic resistance on the compression side of the section.

- Steel profile: $F_{Rd,a,comp} = A_a \times f_{yd,a} = 10600 \times 460 = 4876000$ N.
- Reinforcement top of section: $F_{Rd,s1,comp} = A_{s1} \times f_{yd,s} = 942 \times 435 = 409770$ N.
- Reinforcement two sides:

$$F_{Rd,s2,comp} = 2 \times (L_{p2}/2 - hn) t_{p2} f_{yd,s} = 2 \times (1333/2 - hn) \times 0.764 \times 435$$
$$= 443009 - 665\ hn.$$

- Concrete: net concrete compression area $F_{Rd,c} = A_{c,top} \times f_{cd} =$
- $= [b_w \times (l_w/2 - hn) - A_a - A_{s1} - 2\ t_{p2} \times (L_{p2}/2 - X)] \times f_{cd}$

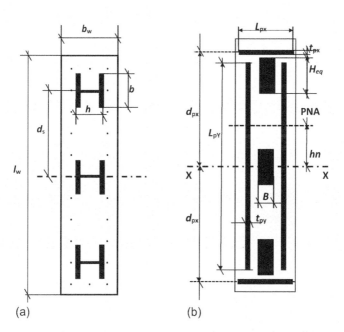

Figure 4.47 a, b Wall section studied. At left, the real section. At right: for simplified analysis

- $= 400 \times (800 - hn) - 10600 - 942 - 2 \times 0.764 \times (1333/2 - hn) = (307439 - 398.5\, hn) \times 28.3$

$F_{Rd,c} = 8700523 - 11277\, hn$.

- Resultant compression resistance:

$F_{Rd,comp,tot} = F_{Rd,a,comp} + F_{Rd,s1,comp} + F_{Rd,s2,comp} + F_{Rd,c}$
$= 4876000 + 409770 + 443009 - 664.7\, hn + 8700523 - 11277\, hn$
$= 14429302 - 10612.3\, hn$.

Plastic resistance on the tension side of the section.

- Steel profiles: $F_{Rd,a,tens} = 2A_a \times f_{yd,a} = 2 \times 10600 \times 460 = 9752000$ N.
- Reinforcement bottom of section: $F_{Rd,s1,tens} = A_{s1} \times f_{yd,s} = 942 \times 435 = 409770$ N.
- Reinforcement two sides:

$F_{Rd,s2,tens} = 2 \times (L_{p2}/2 + hn)t_{p2}\, f_{yd,s} = 2 \times (1333/2 + X) \times 0.764 \times 435$
$= 443009 - 664.7\, hn$.

- Concrete: $F_{Rd,c} = 0$ (tension).
- Resultant tension resistance: $F_{Rd,tens} = F_{Rd,a,tens} + F_{Rd,s1,tens} + F_{Rd,s2,tens}$.
$= 10604779 - 664.7\, hn$

Equilibrium: $F_{Rd,comp,tot} = F_{Rd,tens,tot} \Rightarrow 14429302 - 10612.3\, hn = 10604779 - 664.7\, hn$
$\Rightarrow$ Ordinate hn of PNA: $hn = 339$ mm.

The moment of resistance $M_{pl,Rd}$ of the hybrid section is calculated by adding the moments of the resistance due to the tension and compression forces at any point. Hereunder the moments are calculated about the axis of symmetry of the section:

- Concrete: $F_{Rd,c} = 8700523 - 11277\, hn = 4877620$ N

lever arm: $339 + (800 - 339)/2 = 569.5$ mm
$M_c = 3604$ kNm.

- Steel profiles: $M_a = 2F_{Rd,a,comp} \times d_s = 2 \times 4876 \times 600 = 5851200$ kNmm $= 5851$ kNm.
- Steel profile at mid-section: lever arm = 0 and moment = 0.
- Top and bottom reinforcement: $M_{s1} = 2F_{Rd,s1} \times 750 = 2 \times 410 \times 50 = 615000$ kNmm = 615 kNm.
- Side reinforcement above PNA:

Length of equivalent plate: $L_{p2}/2- hn$ = 1333/2–339= 327.5 mm
Force: 327.5 x t_{p2} x $f_{yd,s}$ = 327.5 x 0.764 x 435 = 108841 N
Lever arm: hn + 327.5/2= 339 + 164 = 503 mm
Moment = 2x 108841 x 503 = 2x54747023 Nmm = 110 kNm.

- Side reinforcement below PNA:

Length of equivalent plate: $L_{p2}/2$ + hn = 1333/2+ 339= 10055 mm.
    Force: 1005.5 x t_{p2} x $f_{yd,s}$ = 1005.5 x 0.764 x 435 = 334167 N.
    ever arm: 1005.5 /2 – hn = 503 – 339 = 164 mm:
    mMoment = 2x 334167 x 164 = 2x54803530 Nmm = 110 kNm:
    => moment of resistance $M_{pl,Rd}$ of the hybrid section in pure bending –
key Point B on the N-M interaction curve: $M_B = M_{pl,Rd}$ = 2778 + 5851 + 615
+ 110 + 110 = 9464 kNm:

Note: some terms in the calculations of the PNA position hn and of $M_{pl,Rd}$
have very low importance. They can be ignored or approximated to shorten
the calculations without impairing significantly accuracy: side reinforcement A_{s2} ignored; concrete gross area considered instead of the net area; no
definition of the equivalent plate for the top and bottom reinforcement A_{s1}
if only bending about X axis is considered.

Check calculations of $M_{pl,Rd}$ as the sum of moments at PNA.
Concrete   0.5(800–339) x 4877620 = 1124 kNm
Profiles   Top profile: (600–339)x4876
                Mid-profile: 339x4876
                Bottom profile: (600+339)x4876
                Total profiles: (600+600+339)4876=7504 kNm
Rebars s1   Top: (750–339)x409770
                Bottom: (750+339) x 409770
                Total: 2 x 750 x 409770 = 614 kNm
Rebars s2   Above PNA 0.5(1333/2 –339) x 108841 = 18 kNm
                Below PNA 0.5(1333/2 +339) x 108841 = 168 kNm
Total: $M_B = M_{pl,Rd}$ = 1124 + 7504 + 614 + 18 +168 = 9428 kNm ≈ 9464 kNm
(The difference is due to rounding).

Point C
$M_C = M_B = M_{pl,Rd}$ = 9464 kNm
$N_C = 0.5 \times 0.85\ A_c\ f_{cd}$ = 0.85 x 28.3 x 604280 = 14535955 N = 14536 kN

Maximum moment of resistance – key Point D (N_D, M_{max})
$N_D = 0.5 \times 085\ A_c\ f_{cd}$ = 0.5 x 0.85 x 28.3 x 604280 = 7267978 N = 7268 kN
$M_{max,Rd} = W_{pa}\ f_{yd,a}$ + 0.5 $W_{pc}\ f_{cd}$ + $W_{ps}\ f_{yd,s}$

where W_{pa}, W_{pc} and W_{ps} are the plastic modulus of steel profiles, of the overall concrete and of the reinforcement about the axis of symmetry of the composite cross-section.

Concrete   Equivalent width for the whole section: 604280/1600 = 378 mm
W_{pc} = 378 x 800 x 400 x 2 = 241920000 mm³
Moment: 0.5 W_{pc} x $0.85f_{cd}$ = 0.5 x 241920000 x 28.3 = 3423 kNm.

Profiles   Top + bottom: W_{pa} = A_s x 600 x 2 = 10600 x 600 x 2
= 12720000 mm³
Mid-section: W_{pa} = 120 x 44.17 x 60 x 2 = 636048 mm³
Moment: $W_{pa} f_{yd,a}$ = (12720000 + 636048)x$f_{yd,a}$
= 13356048 x 460 = 6144 kNm.

Rebars   Top + bottom s1: W_{ps1} = 942 x 750 x 2 = 1413000 mm³
Two sides s2: W_{ps2} = (bh²/4) x2= 0.764 x 1333² /2= 678771 mm³
Moment: $W_{ps} f_{yd,a}$ = 2091771 x 435 = 910 kNm.

Maximum moment of resistance $M_{max,Rd}$ = 3423 + 6144 + 910 = 10477 kNm.

4.11.2 Resistance of a hybrid wall to combined shear, axial force and bending

4.11.2.1 Hypothesis

The wall region for which the calculations hereunder are made is not disturbed by a local introduction of forces or a side reaction: it is a region "B" because the force application at storey levels is made through axially stiff beams which are part of the wall and because the reaction at the base is given by a stiff foundation (see Section 4.8.1).

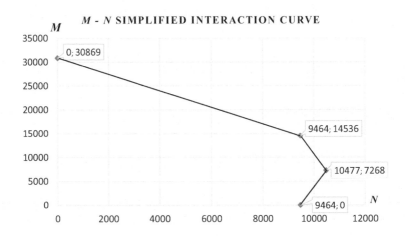

Figure 4.48 M-N interaction curve of the designed section

4.11.2.2 Design action effect

The design action effects at the base of the wall established by the analysis of the structure submitted to the factored loads are:

$$M_{Ed} = 9950 \text{ kNm } N_{Ed} = 9140 \text{ kN } V_{Ed} = 4950 \text{ kN}$$

4.11.2.3 Stiffness K_{RC} of the reinforced concrete truss in the hybrid truss model

K_{RC} is calculated according to Equation (4.25) for diameter 12 transverse reinforcement at a vertical step s =75mm: 2 diameter 12 = 2 x 113 = 226 mm²/stirrup

$\rho_t = A_{sw} / s\, b_w = 226/(75\text{x}400)=0.0075 = 0.75\%$
$m = n_0 = E_s/E_{cm} = 210000/37000 = 5.67$

Factor η for N_i =1 internal profile with a web parallel to the wall's strong bending axis:

$$\eta = 1 + \frac{2N_i\, m b t_f}{b_w z} = 1 + \frac{2 \times 3 \times 5.67 \times 240 \times 17}{400 \times 1200} = 1.29$$

Equation (4.25) gives:

$$K_{RC} = \frac{0.22\, z\eta E_s b_w \rho_t}{0.108\eta + m\rho_t} = \frac{0.22 \times 1200 \times 1.29 \times 210000 \times 400 \times 0.0075}{0.108 \times 1.29 + 5.76 \times 0.0075}$$

$$= 1179 \times 10^6 \text{mm}^{-1}$$

4.11.2.4 Stiffness K_{SP} due to steel profiles in the hybrid truss model

As explained in Section 4.5.5, if the steel profiles are parallel to the long side of the wall, the shear area A_v per profile is found as $A_v = 2bt_f = 2$ x 240 x 17 = 8160 mm²/encased steel profile.

N = 3 encased profiles
$K_{SP} = NG_s A_v = 3$ x 80769 x 8160 = 1977x10⁶ mm⁻¹

4.11.2.5 Shear action effects in the hybrid truss model

Shear action effects $V_{Ed,SP}$ applied to the encased profiles:

$$\frac{K_{SP}}{K_{RC} + K_{SP}} = 0.626 \qquad\qquad \text{(Equation 4.17)}$$

$\Rightarrow V_{Ed,SP} = 0.626 \times 4950 = 3098 kN$ for the total of three encased profiles.

Shear action effects $V_{Ed,RC}$ applied to the reinforced concrete truss:

$$\frac{K_{RC}}{K_{RC} + K_{SP}} = \frac{1179}{1977 + 1179} = 0.373 \qquad \text{(Equation 4.18)}$$

$$\Rightarrow V_{Ed,RC} = 0.373 \times V_{Ed} = 0.373 \times 4950 = 1846\, kN$$

4.11.2.6 Limitation of tension strength of steel profiles due to shear stresses

Design shear action effect per profile: V_{Ed} = 3098/3=1032 kN; shear area A_v: 8160 mm²

$V_{pl,Rd} = A_v f_y/\sqrt{3}$ = 8160 x 460/1.73=2169 kN and $V_{Ed}/V_{pl,Rd}$= 0.475 < 0.5

There is no reduction of the available resistance in tension/compression of the encased profiles due to shear (see Equation (4.39) in Section 4.6.4).

4.11.2.7 Design resistance under combined bending and axial force

Because there is no reduction of the available resistance in the tension of the encased profiles due to shear, an M-N interaction curve defined considering the nominal yield strength f_y of the profiles may be used. With the simplified interaction curve in Section 4.11.1, the moment of resistance for N_{Ed} = 9402 kN is found by linear interpolation between Point C and Point D.

At Point C: $N_{Rd,C}$= 14536 kN; $M_{Rd,C}$ = 9464 kNm
At Point D: $N_{Rd,D}$ = 7268 kN; $M_{Rd,D}$ =10477 kNm
For N_{Ed} = 9140 kN: M_{Rd} =9464+(10477–9464)x[(14536–9140)/(14536–7268)]=9464+752
M_{Rd} =10216 kNm > M_{Ed} = 9950 kNm

4.11.2.8 Design resistance in shear of reinforced concrete limited by crushing of concrete compression struts

$V_{Rd,max} = \alpha_{cw} b_w z \nu_1 f_{cd} /(cot\theta + tan\theta)$ Equation (4.44)

α_{cw} depends σ_{cp}, average stress on an equivalent section of concrete taking into account all bars and steel profiles, with $m = E_s/E_{cm}$ = 210000/37000 = 5.67

Equivalent section of concrete: 5.67x3x10600 + 604280 + 5.67x2x(1018 +942)=806812 mm²
σ_{cp} = 9982000/806812 = 12.4 MPa σ_{cp}/f_{cd} = 12.4/33.3 = 0.37 α_{cw}=1+0.37=1.37
b_w = 400 mm

z is taken as the distance between the two external profiles: $z = 2d_s = 2$ x $600 = 1200$ mm

$$v_1 = 0.6(1 - f_{ck} / 250) = 0,6 \ (1 - 50 / 250) = 0.48$$

$\theta = 35°$ (see Section 4.5.4): $cot\theta = 1.43 \ tan\theta = 0.70 \ cot\theta + tan\theta = 2.13$

$V_{Rd,max} = 1.37$ x 400 x 1200 x 0.48 x $33.3 / 2.13 = 4934000$ N $= 4934$ kN
Design check for resistance of concrete compression struts: $V_{Rd,max} = 4934$ kN$>V_{Ed,RC}=1846$kN

It may be noted that resistance to failure of concrete compression struts $V_{Rd,max}$ is, in this design, equal to V_{Ed} applied to the hybrid section: $V_{Rd,max} = 4934$ kN $\approx V_{Ed} = 4950$ kN

4.11.2.9 Design resistance in shear of reinforced concrete limited by yielding of the transverse reinforcement

$$V_{Rd,s} = \frac{A_{sw}}{s} zf_{ywd}cot\theta \quad f_{ywd} = 500 / 1.15 = 435 \text{ MPa}$$

$A_{sw} = 2$ diameter $12 = 2$ x $113 = 226$ mm^2/stirrup $A_{sw}/s = 226/75 = 3.01$
$V_{Rd,s} = 3.01$ x 1200 x 435 x $1.43 = 2247000$ N $= 2247$ kN
Design check for the resistance of stirrups: $V_{Rd,s} = 2247$ kN $> V_{Ed,RC} = 1846$ kN

It may be noted that thanks to the contribution of the encased profiles to shear resistance, the resistance at yield of the stirrups is significantly smaller than the total shear V_{Ed} applied to the hybrid section: $V_{Rd,s} = 2247$ kN $< V_{Ed} = 4950$ kN.

Check of design stress in stirrups: $F_{Ed,stirrup}=V_{Ed,RC} = 1846$ kN over $zcot\theta = 1716$ mm.

The section area of stirrups over $zcot\theta$ is: $(226$ x $1.716)/0.75 = 5171$ mm^2.
The design stress in stirrups is: $(1846$x$10^3)/5171=356$ MPa $< 500/1.15=434$ MPa.

4.11.2.10 External profile. Design for resistance to longitudinal shear at concrete-profile interface

$A_{prof} = 10600$ mm^2 for one HE240B external profile.
The rebars part of one chord of the truss are 3 diameter 20 and 4 diameter 12.
$A_{chord} = A_{prof} + A_{rebars} = 10600 + 3$ x $314 + 4$ x $113 = 11994$ mm^2.
$\alpha = A_{prof}/A_{chord} = 10600/11994 = 0.88$.
$l_{profile,perim} = 1384$ mm for a HE240B.

The following V_{Ed}, $V_{Ed,l}$ and V_{Rd} are calculated for a wall height equal to $zcot\theta$ = 1716 mm.

$$V_{Ed,l} = 1.43\alpha V_{Ed,RC} = 1.43 \text{ x } 0.88 \text{ x } 1846 = 2323 \text{ kN} \qquad \text{Equation (4.32b)}$$

$$V_{Rd,bond} = 1.43 \ \tau_{Rd} z \ l_{profile,perim} = 1.43\text{x}0.3\text{x}1200\text{x}1384$$
$$=712483 \text{ N}=712 \text{ kN} \qquad \text{Equation (4.43a)}$$

$$V_{Rd,friction} = \mu \ V_{Ed,RC} = 0.5 \text{ x } 1846 = 923 \text{ kN} \qquad \text{Equation (4.43c)}$$

The required number of shear connectors depends on the designer's decision on the friction coefficient μ (taken above as μ=0.5) and on ($\tau_{Rd} x$ $l_{profile,perim}$) adopted – see Section 4.7.2. Hereabove, $V_{Rd,bond}$ and $V_{Rd,friction}$ are calculated above with standard Eurocode values and it results in $V_{Rd,bond}+V_{Rd,friction}$ = 712+923 = 1635 kN.

Connectors are required for a longitudinal shear action effect equal to:

$$V_{Ed,l} - (V_{Rd,bond}+V_{Rd,friction}) = 2323 - 1635 = 688 \text{ kN}$$

A diameter of 16 mm headed studs resistance to shear is P_{Rd} = 64.3 kN.

The number of headed studs required per m is: 688/(1.716 x 64.3)= 6.2 => 7/m.

Note: if no confidence was given to resistance to longitudinal shear by bond and friction, connectors should be placed to resist the total longitudinal shear $V_{Ed,l}$= 2323 kN and 22 studs/m would be required.

4.11.2.11 Internal profile. Design for resistance to longitudinal shear at the concrete-profile interface

The total design action effect $V_{Ed,l}$ at an internal profile is double of $V_{Ed,l}$ at an external profile because it affects the two sides of the profile – see Section 4.7.2). It results in:

$$V_{Ed,l} = 2 \text{ x } 2323 \text{ kN} = 4646 \text{ kN}$$

On the resistance side, $V_{Rd,bond}+V_{Rd,friction}$ are like the above:

$$V_{Rd,bond}+V_{Rd,friction} = 712+923 = 1635 \text{ kN}$$

Shear connectors should resist: 4646 – 1635 = 3011 kN
The number of diameter 16 mm headed studs required per m is:
3011/(1.716 x 64.3)=27.3/m => 28/m, meaning 14 studs/m on each side of the internal profile.

4.11.2.12 Design for ductility

The limitation in applied axial force N_{Ed} of Eurocode 8 (2022) for high ductility of composite walls – Equation (4.49) – is satisfied:

$$v_{dc} = \frac{N_{Ed}}{f_{cd}\left[A_c + n_0\left(A_s + A_a\right)\right]} = \frac{9140}{33.3\left[604280 + 5.67\left(942 + 1018 + 3\times10600\right)\right]} = 0.345$$

$v_{dc} = 0.345 < 0.35$

Earthquake-resistant design aiming at ductility also requires that Equation (4.47b) and (4.47c) be satisfied: $V_{Rd,s} \geq \gamma_{V1} V_{Ed,RC}$ and $V_{Rd,max} \geq \gamma_{V2} V_{Rd,s}$

In the design presented above: $V_{Rd,max}$= 4934 kN, $V_{Rd,s}$ =2247 kN and $V_{Ed,RC}$=1846 kN.

Thus $V_{Rd,max} = 2.19\ V_{Rd,s}$ and $\gamma_{V2} > 1.5$ so that one condition for medium and high ductility (displacement ductility greater than 2 and up to 6) is satisfied.

$V_{Rd,s} = 1.21$, $V_{Ed,RC}$ and $\gamma_{V1} > 1.2$: the design is valid for medium ductility (displacement ductility below 4), not for high. It is however possible to increase the displacement ductility by means of an increase of $V_{Rd,s}$ from $1.21 V_{Ed,RC}$ to $1.5 V_{Ed,RC}$. This can be realized by an increase in the stirrup parameter A_{sw}/sb_w. If the increase in A_{sw}/sb_w is between 24% and 28%, the ratio $V_{Rd,max}/V_{Rd,s}$ is still greater than 1.5 and both of the conditions expressed by Equation (4.47b) and (4.47c) are satisfied with the greatest values of γ_{V1} and γ_{V2}, so that the available displacement ductility is 6 (high ductility).

4.11.2.13 Compliance with limitations of longitudinal steel content

Percentage of steel reinforcement: $\rho_s = A_s/A_{c,gross}$ = (942+1018)/(1600x400) = 0.0031 = 0.31%

Steel profiles contribution ratio: $\delta = A_a\ f_{yd}/\ N_{pl,Rd}$ = (3x10600x460)/ 30869.10^3= 0.47

ρ_s and δ comply with the requirements proposed in Section 4.10.1:

$0.2\% \leq \rho_s = 0.31\% \leq 4\%\ \delta = 0.47 \leq 0.90$

4.11.3 Calculation of longitudinal shear by the classical beam method and comparison

For the sake of a comparison with the longitudinal shear action effect calculated in Section 4.11.2, a full concrete equivalent section of the wall is defined with the classical beam method – see Section 4.4. The reference section is the one in Figure 4.47b. The equivalent section is shown in Figure 4.49. The modular ratio is: $m = n_0 = E_s/E_{cm}$ = 210000/37000 = 5.67.

The first moment of area in sections 1, 2 and 3 and the second moment of area are respectively:

$S_1 = 24.6.10^6$ mm^3 $S_2 = 125.9.10^6$ mm^3 $S_1 = 387.10^6$ mm^3 $I = 2.11.10^{11}$ mm^4

The longitudinal shear in sections 1, 2 and 3 related to the design transverse shear equal to 4950 kN is calculated with Equation (4.11):

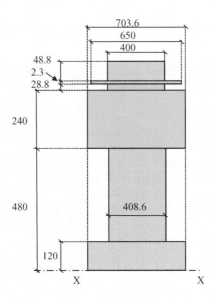

Figure 4.49 Concrete section equivalent to the simplified section at Figure 4.47.

$V_{Ed,l,1}$ = 4950 x 24.6.10^6/2.11.10^{11}= 0.577 kN/mm = 577 kN/m
$V_{Ed,l,2}$ = 4950 x 125.9.10^6/2.11.10^{11}= 2.95 kN/mm = 2950 kN/m
$V_{Ed,l,3}$ = 4950 x 387.10^6/2.11.10^{11}= 4.33 kN/mm = 4330 kN/m
The total longitudinal shear at the concrete-external profile interface is:
$V_{Ed,l}$ = $V_{Ed,l,1}$ + $V_{Ed,l,2}$ = 3527 kN
to be compared to $V_{Ed,l}$ = 2323 kN found by the hybrid truss method.
The total longitudinal shear at the concrete-external profile interface is:
$V_{Ed,l}$ = 2 x 4330 = 8660 kN
to be compared to $V_{Ed,l}$ = 4646 kN found by the hybrid truss method.
The difference in results is explained by the fact that the classical beam method ignores the deformations of sections due to shear and, accordingly, the contribution of the encased steel profiles to the shear stiffness of the section. Therefore the transverse shear, taken by what is considered a classical reinforced concrete beam in the classical beam method, is V_{Ed}=4950kN, while it is $V_{Ed,RC}$=1846 kN in the reinforced concrete part of the hybrid section.

REFERENCES

AISC 341–16 - Seismic provisions for structural steel buildings.
ACI 318-19. Building Code Requirements for Structural Concrete. Commentary on Building Code Requirements for Structural Concrete. American Concrete Institute. ISBN: 978-1-64195-056-5; DOI:10.14359/51716937

Bogdan, T., and Chrzanowski, M. (2018) Mega columns with several reinforced steel profiles – Experimental and numerical investigations. 12th International Conference on Advances in Steel-Concrete Composite Structures (ASCCS 2018). http://dx.doi.org/10.4995/ASCCS2018.2018.7126

Bogdan, T., Chrzanowski, M., and Odenbreit, C. (2019, October) Mega columns with several reinforced steel profiles – Experimental and numerical investigations. Structures, 21, 3–21. https://doi.org/10.1016/j.istruc.2019.06.024

Cho, S. H., Tupper, B., Cook, W. D., and Mitchell, D. (2004) Structural steel boundary elements for ductile concrete walls. https://doi.org/10.1061/(ASCE)0733-9445(2004)130:5(762)

Chrzanowski, M. (2019) Shear transfer in heavy steel-concrete composite columns with multiple encased steel profiles. PhD thesis. FSTC-2019-40. Luxemburg University website.

Corley, W. G., Fiorato, A. E., and Oesterle, R. G. (1981) *Structural Walls*, Special Publication 72–4, American Concrete Institute, 77–131.

Dan, D., Fabian, A., and Stoian, V. (2011a) Nonlinear behavior of composite shear walls with vertical steel encased profiles. *Engineering Structures*, 33(10), 2794–2804. https://doi.org/10.1016/j.engstruct.2011.06.004

Dan, D., Fabian, A., and Stoian, V. (2011b). Theoretical and experimental study on composite steel–concrete shear walls with vertical steel encased profiles. *Journal of Constructional Steel Research*, 67(5), 800–813. https://doi.org/10.1016/j.jcsr.2010.12.013

Degee, H., Plumier, A., Mihailov, B., Dragan, D., Bogdan, T., Popa, N., De Bel, J. M., Mengeot, P., Hjiaj, M., Nguyen, Q. H., Somja, H., Elghazouli, A., and Bompa, D. (2017) Smart composite components concrete structures reinforced by steel profiles (SmartCoCo): Final report, EUR 28914 EN. https://doi.org/10.2777/587887

Elwood, K. J., and Eberhard, M. O. (2006). Effective Stiffness of Reinforced Concrete Columns, PEER Research Digest No. 2006-1, A publication of the Pacific Earthquake Engineering Research Center. https://apps.peer.berkeley.edu/publications/research_digest_2006/rd2006-1.pdf

Epackachi, S., Whittaker, A., Huang, Y. N. (2015) Analytical modeling of rectangular SC wall panels. *Journal of Constructional Steel Research* 105, 49–59. http://dx.doi.org/10.1016/j.jcsr.2014.10.016

Eurocode 2. (2004) Design of concrete structures — Part 1–1: General rules and rules for buildings, EN 1992-1-1. CEN European Committee for Standardization, Brussels.

Eurocode 2. (2022) Design of concrete structures—General rules and rules for buildings/ prEN 1992-1-1:2022. CEN European Committee for Standardization, Brussels.

Eurocode 3. (2021) EN 1993-1-1. Design of steel structures - Part 1–1: General rules and rules for buildings.. CEN European Committee for Standardization, Brussels.

Eurocode 4. (2004) Design of composite steel and concrete structures - Part 1–1: General Rules for buildings, EN 1994-1.1. CEN European Committee for Standardization, Brussels.

Eurocode 8. (2004) EN1998-1. Design of structures for earthquake resistance. Part 1: General rules, seismic action and rules for buildings. CEN European Committee for Standardization, Brussels.

212 Design of hybrid structures

XXEurocEurocode 8. (2022) EN1998-1-1. Design of structures for earthquake resistance. Part 1-1: General rules, seismic action. CEN European Committee for Standardization, Brussels.

FEMA 356. (2000) Prestandard and commentary for the Seismic rehabilitation of buildings. Federal emergency management agency, United States.

Gerardy, J. C., Plumier, A., Bogdan, T., and Degee, H. (2013) Design example of a column with 4 encased steel profiles. Conference: 10th Pacific Structural Steel Conference (PSSC 2013). https://doi.org/10.3850/978-981-07-7137-9_003; https://www.researchgate.net/publication/269207314

Grandić, D., Šćulac, P., and Štimac Grandić, I. (2015) Shear resistance of reinforced concrete beams in dependence on concrete strength in compressive struts, Tehnicki Vjesnik August 2015. https://doi.org/10.17559/TV -20140708125658

INSA, (2017) SMARTCOCO Project, Deliverable D7.2 Test Report (Internal report, available on request at a.plumier@uliege.be)

Ji, X., Quian, J., and Jiang, Z. (2010) Seismic behaviour of steel tube reinforced concrete, steel & composite structures—Proceedings of the 4th international conference, copyright 2010 ICSCS organisers. Published by Research Publishing. https://doi.org/10.3850/978-981-08-6218-3 CC-We012

Ji, X., Sun, Y., Quian, J., and Lu, X. (2015) Seismic behavior and modeling of steel reinforced concrete (SRC) walls. *Earthquake Engineering & Structural Dynamics*. https://doi.org/10.1002/eqe.2494

Lai, M. H., and Ho, J. C. M. (2017, October) An analysis-based model for axially loaded circular CFST columns, 2017. Thin-Walled Structures, 119, 770–781. https://doi.org/10.1016/j.tws.2017.07.024

Moehle, J. (2015) Seismic design of reinforced concrete buildings. McGraw-Hill Education. ISBN: 978-0-07-183944-0, MHID: 0-07-183944-5

Nethercot, D. (2004) *Composite construction*. Spon Press – Taylor and Francis. CRC Press DOI https://doi.org/10.4324/9780203451663

Plumier, A. (2015) A design method for walls with several encased steel profiles. 11th International Conference on Advances in Steel and Concrete Composite Structures. Tsinghua University. Beijing, China, December 3–5, 2015.

Plumier, A., Bogdan, T., and Degee, H. (2012) Design of columns with several encased steel profiles for combined compression and bending. Final report for ArcelorMittal. Download at https://www.techylib.com/en/view/choruspillow/

Plumier, A., Dragan, D., Huy, N. Q., and Degée, H. (2016) An analytical design method for steel-concrete hybrid walls. *Structures*. http://dx.doi.org/10.1016 /j.istruc.2016.12.007

Priestley, M., and Kowalsky, M. (1998) June, Aspects of drift and ductility capacity of rectangular structural walls. *Bulletin of the New Zealand Society for Earthquake Engineering*. http://dx.doi.org/10.5459/bnzsee.31.2.73-85

Qian, J., Jiang, Z., and Ji, X. (2012, March) Behavior of steel tube-reinforced concrete composite walls subjected to high axial force and cyclic loading. Engineering Structures, 36, 173–184. https://doi.org/10.1016/j.engstruct .2011.10.026

Roik, K., and Bergmann, R. (1992) Composite columns. In P. J. Dowling, J. E. Harding, and R. Bjorhovde (eds.), *Constructional steel design: An international guide*, 443–470. Elsevier Science Publishers Ltd., London and New York. DOIhttps://doi.org/10.1201/9781482296709

SMARTCOCO. (2017) EUR 28914 EN, smart composite components - Concrete structures reinforced by steel profiles, European commission. Research programme of the research funds for coal and steel, TGS8 2016. RFSR-CT-2012-00031. ISBN 978-92-79-77016-6. https://doi.org/10.2777 /587887; https://op.europa.eu/en/publication-detail/-/publication/aee33c6b -58b6-11e8-ab41-01aa75ed71a1

Sezen, H., and Moehle, J. P. (2004) Shear strength models for lightly reinforced concrete columns. *Journal of Structural Engineering*, 130(11), 1692–1703.

St Venant, Adhemar Jean Claude Barre (1844) Mémoire sur la résistance des solides. https://play.google.com/store/books/details

Timoshenko, S., and Goodier, J.N. (1951) Theory of Elasticity, Second Edition, McGraw-Hill Book C., Inc., 1951.

Todea, V., Dan, D., Florut, S. C., and Stoian, V. (2021) Experimental investigations on the seismic behavior of composite steel concrete coupled shear walls with central openings. *Structures*. 2021(33), 878–896.

TTTrabuco, D., Gerardy, J.C., Xiao, C., Davies, D. (2016) Composite Megacolumns: Testing Multiple Concrete-Encased Hot-Rolled Steel Sections. A Council of Tall Buildings and Urban Habitat and ArcelorMittal Publication. ISBN: 978-0939493-53-1.

Tupper, B. (1999) Seismic response of reinforced concrete walls with steel boundary elements, PhD Thesis, McGill University, Montréal, Canada, www.nlc-bnc.ca/obj/s4/f2/dsk1/tape8/PQDD_0023/MQ50667.pdf

Varma, A., Zhang, K., Chi, H., Booth, P., and Tod Baker, T. (2011) In-plane shear behavior of SC compositewalls: Theory vs. experiment, transactions, SMiRT 21, 6-11 November, 2011, New Delhi, India.

Wu, L., Tiang, Y., Su, Y., and Chen, H. (2018) Seismic performance of precast composite shear walls reinforced by concrete-filled steel tubes. Engineering Structures, 162, 72–83. https://doi.org/10.1016/j.engstruct.2018.01.069

Zhang, J., Li, X., Cao, W., and Yu, C. (2020) Seismic behavior of composite shear walls incorporating high-strength materials and CFST boundary elements, October 2020. *Engineering Structures* 220, 110994. https://doi.org/10.1016 /j.engstruct.2020.110994

Zhou, Y., Lu, X. L., and Dong, Y. (2010) Seismic behaviour of composite shear walls with multi embedded steel sections. Part 1: Experiment. *The Structural Design of Tall and Special Buildings*, 19, 618–636. https://doi.org/10.1002/ tal.597

Zhou, J., Fang, X., and Jiang, Y. (2021) Cyclic behaviour of concrete encased high strength concrete filled steel tube composite walls: An experiment. *Structural Concrete*, 22, 691–708. https://doi.org/10.1002/suco.201900233

Chapter 5

Transition between composite and reinforced concrete components

Hervé Degée and Rajarshi Das

CONTENTS

5.1 INTRODUCTION

The goal of this chapter is to suggest design options ensuring a proper transition between structural components in non-conventional steel-concrete structures, that are not covered – or only in a limited way – by the current design standards.

The suggested solutions are in principle applicable to various structural assemblies and have been pushed forward by some specific needs as

DOI: 10.1201/9781003149811-5

described in Chapter 1. It covers for instance the connection of steel or steel-concrete composite beams to reinforced concrete columns or walls, or the transition zone in columns that are partly reinforced concrete and partly composite.

Among those types of connections, only one situation is explicitly covered by standards based on extensive research work, namely the connection of a steel or composite coupling beam to a reinforced concrete wall, relying on a sufficient length of the steel part of the beam embedded inside the concrete and adequate longitudinal and transverse reinforcement, as illustrated in Figure 5.1 and explained in Section 5.2.

The following sections present then a design procedure for two other situations:

- An alternative solution for the connection of coupling beams not resorting to the embedded length but to an embedded vertical steel component to which the horizontal steel component is connected by a moment connection, without using any additional shear studs, as illustrated in Figure 5.2.
- The transition zone in columns which are in part made of classical reinforced concrete and in part steel-concrete composite column (see Figure 5.3), with the need to properly transfer bending moment, shear force and normal force from one part to the other. Such design is for instance used in situations where one or more given levels of a building need to be strengthened without changing the outer dimensions of the columns.

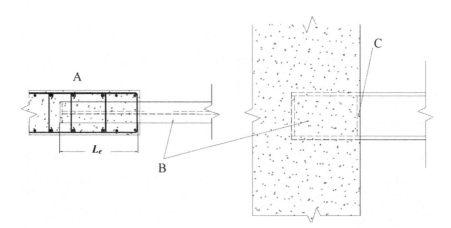

Figure 5.1 Connection of a steel beam B to a wall or a column by means of an embedded length of the framing component with A, additional reinforcement at embedment, and C, face-bearing plate

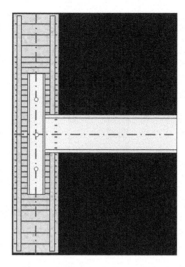

Figure 5.2 Connection of a steel beam to a wall or a column by means of a connection with an embedded vertical steel component

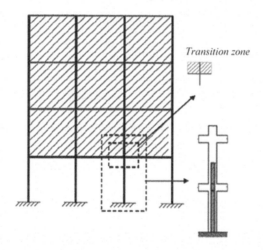

Figure 5.3 Hybrid column partly made of reinforced concrete and partly steel-concrete composite

5.2 JOINT BETWEEN A STEEL OR COMPOSITE COUPLING BEAM AND A REINFORCED CONCRETE WALL RELYING ON THE EMBEDDED LENGTH

5.2.1 Scope

The design of joints between a steel or composite beam and a reinforced concrete wall is of importance for beams coupling two reinforced concrete walls

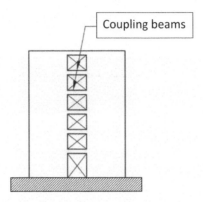

Figure 5.4 Shear walls and coupling beams

as shown in Figure 5.4 in order to make them work together and activate a frame action providing stiffness and strength greater than those of independent walls. Because coupling beams are rather short-span components, their design is often governed by shear, which can make a steel or composite beam solution preferable because it provides easy resistance and ductility under such action effects. In comparison to reinforced concrete, steel or composite coupling beams offer a possibility of beam depth reduction leading to a reduction of storey height and finally of the cost of a building.

Section 5.2 covers the design of joints of steel or composite coupling beams framing into a reinforced concrete wall with or without reinforced concrete or steel boundary elements and in which vertical reactions provided by concrete along the embedded length of the steel or composite beam as shown at Figure 5.5 are the only reactions used to transfer into the wall the bending moment M_{Ed} and the shear V_{Ed} present in the beam at the face of the wall.

Alternative designs such that horizontal reactions are mobilized at the levels of the coupling beam flanges do exist, for instance in the form of headed studs welded onto the beam flanges with horizontal rebars passing around them anchored in a region further to the embedment length. This may also be realized by transmitting the coupling forces through a moment connection with a steel column at the end of the embedment. This design does not correspond to the hypothesis of Section 5.2. They are presented in Section 5.3.

5.2.2 Definition of the required embedment length

There is ample research material concerning coupling beams and there are also code provisions, in particular in AISC341-16 and in the current Eurocode 8 (2004) and its ongoing revision.

The bearing capacity of beams embedded in walls depends on their embedment length L_e which allows for more or less great lever arm of internal forces

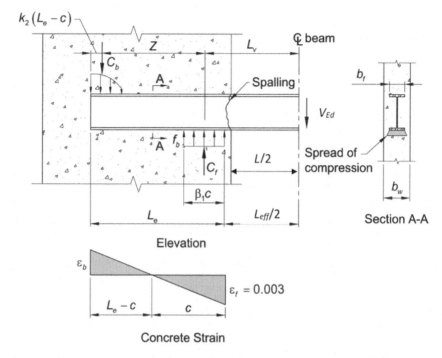

Figure 5.5 Embedment model coherent with Equation (5.2). Courtesy of AISC 341-16

Z as shown in Figure 5.5. Several models have been successively developed, most considering a stress distribution in the concrete related to concrete strain varying linearly up to a maximum of 0.003 close to the face of the wall; we can cite models by Marcakis and Mitchell (1980), Mattock and Gaafar (1982), Park et al. (1982) and Minami (1985). All these models express the strength of the joint in terms of resistance to an applied shear V_{Ed} and of the span L of the coupling beam, which implicitly includes the applied moment M_{Ed} since the shear V_{Ed} in coupling beams due to the frame behaviour is:

$$V_{Ed} = \frac{2M_{Ed}}{L} \tag{5.1}$$

The model on which most design codes are presently based is the one of Park et al. (2005), which corresponds to the set of forces shown in Figure 5.5 and which gives, with Eurocode notations, the design strength V_{Rd} of the connection expressed by Equation (5.2):

$$V_{Rd} = 0.9 \times 40.4\sqrt{f_{cd}} \left(\frac{h_w}{b_f}\right)^{0.66} \beta_1 b_f L_e \left(\frac{0.58 - 0.22\beta_1}{0.88 + \frac{L}{2L_e}}\right) \tag{5.2}$$

where L_e is the embedment length of the beam considered to begin inside the first layer of confining reinforcement nearest to the edge face of the wall, L is the coupling beam clear span, b_w is the thickness of the wall, b_f is the beam flange width, f_{cd} is the design strength of concrete and β_1 is a factor relating the depth of the equivalent rectangular compressive stress block to the neutral axis depth; β_1 may be taken equal to 0.85 for concrete up to C30, equal to 0.65 above C55 and found by interpolation between 0.85 and 0.65 for concrete between C30 and C55.

The minimum calculated embedment length L_e is such that Equation (5.2) is satisfied for a strength V_{Rd} equal to the action effect V_{Ed} at the face of the wall, but the embedment length L_e should always be greater than $1.5d$ where d is the steel beam depth.

Equation (5.2) reflects the local equilibrium at the joint due to the frame effect alone. The local action effects in the joint come in addition to the general stresses in the wall defined by the analysis of the structure. These local action effects around the embedment must be resisted by longitudinal (= vertical) and transverse (= horizontal) reinforcement of the wall, which is additional to those required from the action effects established by the global analysis of the structure.

5.2.3 Design for static and low ductility applications

If it is not intended to develop the plastic capacity of the coupling beam, like under wind action or for low ductility design ("ordinary" shear walls in AISC, DCL design in Eurocode 8), V_{Ed} is established by the analysis of the structure and it is permitted to use vertical reinforcement placed for other purposes, such as for vertical boundary components, as part of the required vertical reinforcement. The necessary reinforcement may be found by calculating the reaction forces C_f and C_b shown in Figure 5.5 and placing reinforcement accordingly, but for static and low ductility applications, a simplified rule is given in AISC341-16: vertical wall reinforcement with a total design axial resistance equal to V_{Ed} should be present over the embedment length of the beam, with two-thirds of them located over the first half of the embedment length closest to the wall face. This wall reinforcement should extend vertically on a distance of at least one tension development length, both above and below the flanges of the coupling beam so that it can match the reversal of moment and shear observed under actions like wind or earthquakes.

Horizontal reinforcement, in the form of hoops or cross-ties joining vertical reinforcement present in the long sides of the wall, should be placed to resist the horizontal outward thrust resulting from the spread of the reaction force at the steel flange-concrete interface sketched in section A-A of Figure 5.5. The horizontal reinforcements shown in Figure 5.1 should be present above the upper flange and below the lower flange of the steel profile.

The vertical spacing s of confining hoops should not exceed $s = \min(b_o/2, 260, 9\,d_{bL})$ (in mm) where b_o is the minimum dimension of the concrete core to the centreline of the hoops and d_{bL} is the minimum diameter of the longitudinal rebars. The diameter of the hoops d_{bw} should be at least $d_{bw} = 6$ mm. The distance between consecutive longitudinal (vertical) bars restrained by hoop bends or cross-ties should not exceed 250 mm. It is also advised that at least one vertical line of cross-ties at the end of the embedment length is placed, to prevent the splitting of the wall at that location.

5.2.4 Design for ductility

If it is intended to develop the plastic capacity of the coupling beam, like for "special" shear walls in AISC, or DCM/DCH ductility classes in Eurocode 8, the design shear $V_{Ed,E}$ in the joint due to seismic action should be taken as corresponding to a situation in which plastic hinges are formed at both ends of the coupling beam, as shown at Figure 5.6, so that $V_{Ed,E}$ should be capacity designed to the beam real plastic strength which considers the real yield strength and the strain hardening of the beam materials, as expressed by Equation (5.3):

$$V_{Ed,E} = \frac{2M_{pl,real}}{L}; \; M_{pl,real} = \omega_{sh}\omega_{rm}M_{pl,Rd} \tag{5.3}$$

where ω_{sh} is a coefficient taking into account strain hardening, ω_{rm} a coefficient taking into account the randomness of yield stress and $M_{pl,Rd}$ the design plastic moment of the coupling beam calculated with the design yield stress f_y of the steel beam.

The encased steel section in the embedment length should be wide flange sections.

The embedded steel components should be provided with two regions of vertical transfer reinforcement, each region extending over half the embedment length. The necessary reinforcement should be found by calculating the reaction forces C_f and C_b shown in Figure 5.5 and placing reinforcement

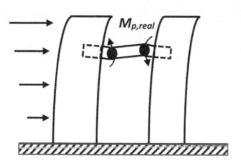

Figure 5.6 Plastic mechanism in coupling beams

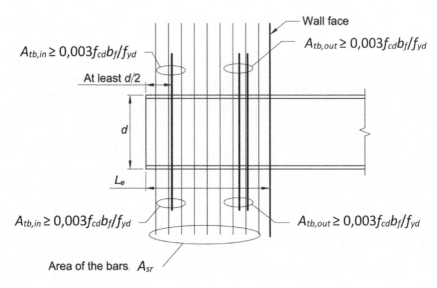

Figure 5.7 Transfer reinforcement in joints of coupling beams designed for ductility.
Courtesy of AISC 341-16.

$A_{tb,in}$ on the inner side and $A_{tb,out}$ on the outer side accordingly. Those vertical transfer reinforcements should be attached to both the top and bottom flanges of the embedded component. As shown in Figure 5.7, the first region should coincide with the location of longitudinal wall reinforcing bars closest to the face of the wall; the reinforcement in the second region should be placed at a distance not less than half the steel beam depth $d/2$ from the termination of the embedment length.

The area $A_{tb,in}$ and $A_{tb,out}$ of the vertical transfer reinforcement should satisfy (5.4):

$$A_{tb,in} \; and \; A_{rb,out} \geq 0.03 f_{ed} L_e \frac{b_f}{f_{yd}} \qquad (5.4)$$

where $A_{tb,in}$ and $A_{tb,out}$ represent respectively the area of transfer reinforcement required in the inner region and in the outer region, f_{yd} is the yield stress of transfer reinforcement, b_f is the width of beam flange and f_{cd} is the design strength of concrete.

The area of vertical transfer reinforcement should not exceed that calculated with (5.5):

$$\left(A_{tb,in} + A_{tb,out}\right) < 0.08 L_e b_w - A_{sr} \qquad (5.5)$$

where A_{sr} is the area of longitudinal wall reinforcement provided over the embedment length L_e and b_w is the width of wall (all in mm²).

Although the required embedment length of the coupling beam may be reduced if the contribution of the transfer reinforcement is considered by adding a second term in Equation (5.2) as proposed in Park et al. (2005), this contribution is generally ignored to avoid excessive inelastic damage in the connection region.

All transfer reinforcement should have at least one full development length starting where they engage the coupling beam flanges. Straight, hooked or mechanical anchorage or headed bars may provide the development. The vertical transfer reinforcement may be attached directly to the top and bottom flanges or be passed through holes in the flanges and be mechanically anchored by bolting or welding. It is permitted to use mechanical couplers welded to the flanges to attach the bars. U-bar hairpin reinforcement as shown in Figure 5.8a may also be used. These hairpins should be alternated to engage the top and bottom flanges.

In order to prevent congestion, the sum of the areas of transfer bars and wall longitudinal bars over the embedment length should not be more than 8% of a section equal to the wall width times the embedment length.

Face-bearing plates, which are full-width stiffeners located on both sides of the web of the coupling beam, should be provided in the steel beam section at the face of the wall. Face-bearing plates realize a confinement of the concrete and contribute to the transfer of forces from the beam to the concrete through direct bearing. Face-bearing plates should be designed as a stiffener at the end of a link beam, with the procedure used in eccentrically braced frames. If it is convenient for formwork, face-bearing plates may be extended beyond the flanges of the coupling beam. It may be required for the shear resistance of the steel beam to also weld stiffeners near the end of the embedded region; if so, these should be aligned with the vertical transfer bars attached to the flanges near the end of the embedded region as shown in Figure 5.8b.

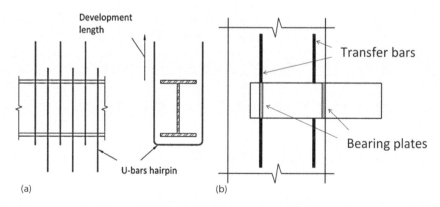

(a)   (b)

Figure 5.8 (a) Alternate U-shaped hairpins as transfer bars. (b) Position of transfer bars and bearing plates in Special Shear Walls of AISC 341-16. Courtesy of AISC 341-16

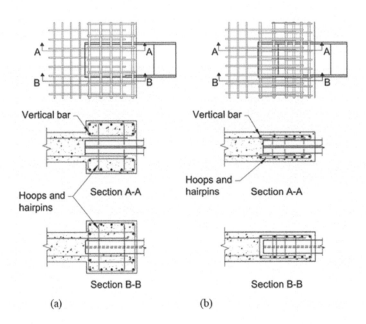

Figure 5.9 a, b Design of reinforcement in a wall boundary zone without bars passing through holes in the web of the steel section: (a) in a "barbell" wall; (b) in a rectangular wall. Courtesy of AISC 341-16

In walls designed for ductility, as explained in Chapter 4, horizontal reinforcement in the form of hoops and cross-ties is required in the boundary zones. The embedment length of coupling beams intersects such boundary zones and transverse reinforcing bars have to pass through the web of the embedded coupling beam. There are alternatives, for instance, bars welded on both sides of the face-bearing plate which participate to complete hoops. Another practical suggested in AISC 341-16 is to place hooked ties on either side of the web and to provide short vertical bars between the flanges to anchor these ties, as shown in Figure 5.9; this solution interrupts the confinement of the complete width of the boundary zone over the height of the embedded steel profile and replaces it with two confined volumes on the sides of the profile. The guidance concerning horizontal reinforcement placed for confinement given in the previous paragraph also apply here.

5.2.5 Resistance of the embedded part of the steel profile

The embedded steel or composite coupling beam is submitted to bending and shear. Within the scope of Section 5.2, in which the only considered reaction forces are vertical at the flange-concrete interface, the bending moment applied to the steel profile in the joint is lower than in the coupling beam span, down to zero at the end of the embedment, and it does not need further consideration. On the contrary, shear in the steel profile may be high over the embedment length. With the notations L_e, c, β_1 and k_2 as defined in Figure 5.5 and taking $c/L_e = 0.66$ and $k_2 = 0.36$, as recommended by Mattock and Gaafar (1982), and $\beta_1 = 0.85$ for a C30 concrete, we obtain a lever arm Z of the reaction forces C_f and C_b: $Z = 0,6L_e$. The equilibrium of the forces shown in Figure 5.10 gives the resultant compression force C_f of the equivalent rectangular compression stress block – Equation (5.6) – and the shear $V_{Ed,panel}$ in the panel zone – Equation (5.7):

$$C_f = V_{Ed} \times \left(\frac{L_{eff}}{2 \times 0.6L_e} \right) \tag{5.6}$$

$$V_{Ed,panel} = C_b = C_f - V_{Ed} \tag{5.7}$$

C_f is $(L_{eff}/1.2L_e)$ times greater than V_{Ed}, and $V_{Ed,panel}$ may also be greater than V_{Ed}. The design check is often more required in the embedded panel zone than in the clear beam span.

The resistance to the applied panel shear $V_{Ed,panel}$ is potentially provided by a combination of the three mechanisms shown in Figure 5.11: the resistance to shear V_{sd} of the steel web panel shown in Figure 5.11a; the resistance V_{csd}

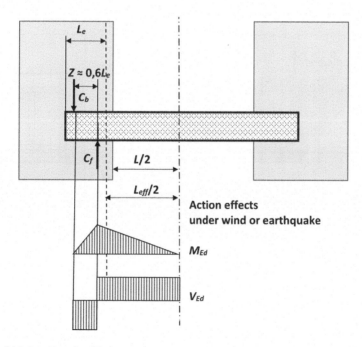

Figure 5.10 Equilibrium of forces giving C_f and $V_{Ed,panel}$

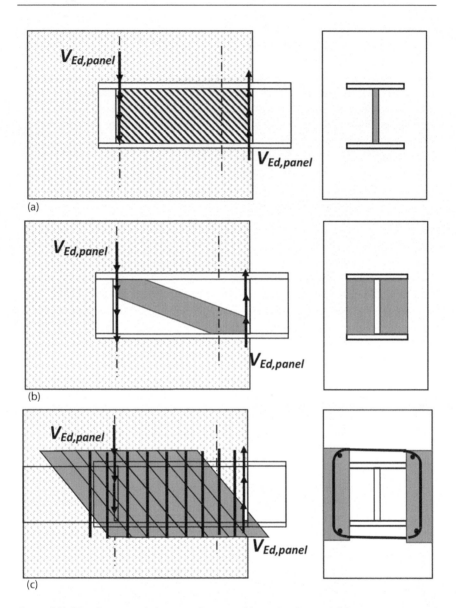

Figure 5.11 The three panel shear mechanisms: (a) steel web panel; (b) concrete encased by the flanges; (c) outer concrete

due to a concrete compression strut developing in the steel panel encased by stiffeners as shown in Figure 5.11b; the resistance to shear V_{cfd} of concrete outside of that encasement shown in Figure 5.11c. The relative contribution of each mechanism depends upon the detailing: presence and dimensions of the stiffeners and shear reinforcement outside of the encasement.

The steel web panel acts like in structural steel connections; it is idealized as carrying pure shear stress over an effective panel length, which depends on the location and distribution of vertical bearing stresses. The concrete compression strut in the encased panel is similar to those used to model shear in composite frame connections. Outside of the concrete panel encased by the stiffeners and the steel coupling beam flanges, compression struts may act with vertical reinforcement to form a truss mechanism the resistance of which may be calculated with the standard equation of reinforced concrete beams. The design shear resistance of a composite joint may be calculated as the sum of the design shear resistance V_{Rd}, V_{csd} and V_{cfd}.

The design shear resistance V_{Rd} of the steel panel may be calculated with Equation (5.8):

$$V_{Rd} = \frac{(f_{y,w} / \sqrt{3})t_w h_w}{\gamma_M} \tag{5.8}$$

where $f_{y,w}$ is the yield strength of the web of the steel section, t_w its thickness, h_w its height and γ_M the material partial factor of Eurocode 3 (2005).

The design shear resistance V_{csd} of the concrete compressive strut (see Figure 5.11b) may be calculated with Equation (5.9).

$$V_{csd} = 0.5 f_{cd} b_p L_p \frac{h_w^2}{h_w^2 + L_p^2} \tag{5.9}$$

$$b_p \le b_f + 5t_p \le 1.5b_f \tag{5.10}$$

where L_p is the length of the web panel between two consecutive stiffeners, the face-bearing plate being one of them; b_p is the effective face-bearing plate width, with limitations given by Equation (5.10).

The design shear resistance V_{cfd} of the concrete compressive field (see Figure 5.11c) may be calculated with Equation (5.11).

$$V_{cfd} = V_s = \frac{f_{ysvd} A_{sv} L_e}{s} \le 0.5 f_{cd} b_o h_w \frac{h_c^2}{h_c^2 + h_w^2} \tag{5.11}$$

where V_s is the design shear resistance due to vertical ties within the beam depth, f_{yshd} is the design yield strength of the reinforcement and A_{sv} is the total cross-section area of ties spaced at s through the web length L_e.

These three mechanisms should deform altogether like a single solid. The problem is automatically solved for the concrete compression strut in the encased panel which is forced to follow the displacement of the steel panel, but to force the concrete outside of the volume encased by the stiffeners and the steel beam flanges requires specific design measures which can be horizontal shear connectors or bars crossing the vertical interface between concrete encased between the steel beam flanges and concrete outside of this encasement. Section A-A of Figure 5.9 shows horizontal hoops and hairpins placed with that intention.

5.3 CONNECTION OF A STEEL OR COMPOSITE BEAM TO A REINFORCED CONCRETE COLUMN OR WALL WITHOUT SUFFICIENT EMBEDDED LENGTH

In those cases where the dimensions of the framing beam and of the reinforced concrete wall or column don't allow a connection solution like described in the previous section, an alternative solution resorting to a vertical piece of steel profile connected rigidly to the framing beam and embedded in the concrete element, as illustrated in Figure 5.2, can become an interesting alternative. The main point is then to ensure a proper load transfer from the embedded vertical element to the surrounding concrete. Such a solution is not really covered by any explicit design method available in the current design standards or other relevant technical references. It is however possible to resort to specific code rules normally applicable for other situations and extend their scope for this design. This approach has been followed for practical cases described in Chapter 1 of this book, as well as for designing test specimens that have been tested in the context of the SMARTCOCO project to validate its efficiency.

This section starts with a description of the suggested design procedure, followed by a summary of the experimental results validating its application. A detailed design example is then given in Section 5.5.1.

5.3.1 Design procedure

The design action effects at the intersection of the beam and column axis (Figure 5.12) are M_{Ed}, V_{Ed} and N_{Ed}. If the beam is a composite steel-concrete element, M_{Ed}, V_{Ed} and N_{Ed} can be subdivided into a part taken by the steel section (index a) and a part taken by reinforced concrete (index c):

$$M_{Ed} = M_{Ed,a} + M_{Ed,c} \qquad\qquad (5.12a)$$

$$V_{Ed} = V_{Ed,a} + V_{Ed,c} \qquad\qquad (5.12b)$$

$$N_{Ed} = N_{Ed,a} + N_{Ed,c} \qquad\qquad (5.12c)$$

Only the parts of M_{Ed}, V_{Ed} and N_{Ed} taken by the steel section, respectively $M_{Ed,a}$, $V_{Ed,a}$ and $N_{Ed,a}$, require specific calculations for their transmission to concrete. The transmission of the "concrete part" $M_{Ed,c}$, $V_{Ed,c}$ and $N_{Ed,c}$ should be made in a classical way.

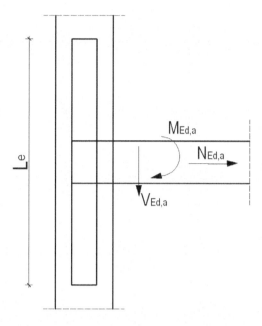

Figure 5.12 Design action effects at beam-column-encased steel profile connection

Eurocode 4 (Section 6.7.3.2 of the 2004 version) indicates how to calculate $M_{Ed,a}$, $V_{Ed,a}$ and $N_{Ed,a}$:

$$M_{Ed,a} = M_{Ed} \times \left(\frac{M_{pl,Rd,a}}{M_{pl,Rd}} \right) \tag{5.13a}$$

where $M_{pl,Rd}$ is the plastic resistance moment of the composite section and $M_{pl,Rd,a}$ is the plastic resistance moment of the steel section.

$$M_{Ed,c} = M_{Ed} - M_{Ed,a} \tag{5.13b}$$

$$V_{Ed,a} = V_{Ed} \times \left(\frac{M_{pl,Rd,a}}{M_{pl,Rd}} \right) \tag{5.14a}$$

$$V_{Ed,c} = V_{Ed} - V_{Ed,a} \tag{5.14b}$$

$$N_{Ed,a} = N_{Ed} \times \left(\frac{A_a f_{yd}}{N_{pl,Rd}} \right) \tag{5.15a}$$

where $N_{pl,Rd} = A_a f_{yd} + 0.85 A_c f_{cd} + A_s f_{sd}$

$$N_{Ed,c} = N_{Ed} - N_{Ed,a} \tag{5.15b}$$

The design should be such that $M_{Ed,a}$, $V_{Ed,a}$ and $N_{Ed,a}$ be transferred properly to the concrete of the column. The vertical encased steel element, which is rigidly connected to the steel beam, should be stiff enough to distribute the reaction forces. The length L_e should comply with:

$L_e / h \geq 6$ and $L_e / h_b \geq 3$; where h is the height of the embedded profile section for the bending axis on which $M_{Ed,a}$ is applied and h_b is the steel beam height.

The final dimensions L_e of the vertical encased steel profile are determined in order to comply with all aspects of the design, in particular the control of the shear resistance in the effective concrete section in the connection zone, i.e. considering only a limited width of the concrete $b_{eff} = b_c - b_a$ where b_a is the width of the vertical encased steel section. It might require increasing L_e in order to obtain a level of shear in the connection zone which is acceptable both at the concrete side (no crushing of compression struts in the effective section) and at the steel side (bars diameter, dimensions and step of stirrups).

The action effects induced by $M_{Ed,a}$ in the various components of the system and which should be considered for design checks are shown in Figure 5.13, where z is the lever arm of internal forces in the beam profile, M is $M_{Ed,a}$, L_e is the length of the profile embedded in the reinforced concrete column and L is the distance between points of zero-moment in the column. A linear distribution of the contact pressure p between concrete and embedded profile is assumed, as illustrated in Figure 5.13b. The action effects induced in the column by $N_{Ed,a}$ and which should be considered for design checks are shown in Figure 5.14. The beam shear $V_{Ed,a}$ induces an axial force in the column-encased steel profile. This force can be transmitted to the concrete of the column along the connection length L_e by bond, shear connectors, end plates and stiffeners or by a combination of those transfer means. The concrete section providing reaction forces in the connection length L_e is submitted to design bending moment M_{Ed} and design shear V_{Ed} due to $M_{Ed,a}$ and $N_{Ed,a}$. Those connection effects are additional to M_{Ed} and V_{Ed} established by the global analysis of the structure.

The resistance of the hybrid joint is conditioned by the resistance of two components (Figure 5.15), i.e. the inner and outer parts of the joint, and by the contact pressure between the embedded profile and concrete. The profile in the inner joint region should be checked in bending and shear and for transmission of the beam shear $V_{Ed,a}$. The outer region should be checked in bending and shear considering a concrete section width $b_{eff} = b_c - b_a$ where b_a is the width of the vertical encased steel section. The contact pressure between the embedded profile and concrete should remain low enough to allow a feasible strut-and-tie mechanism.

The position of the longitudinal and transverse reinforcement should be such that they provide support to the compression struts transmitting the design bending moment M_{Ed} and design shear V_{Ed} due to $M_{Ed,a}$ and $N_{Ed,a}$. Part of or all bents should be facing the compression struts in order to constitute a "suspension" reinforcement bringing the contact force to the other

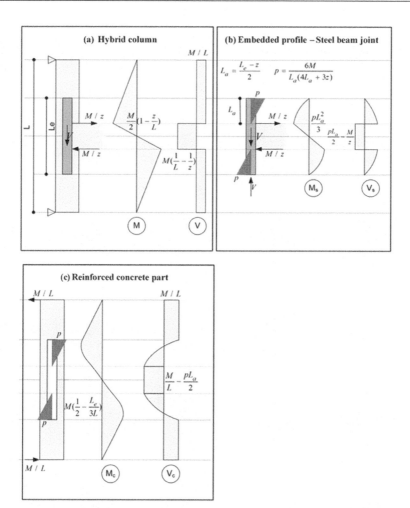

Figure 5.13 a, b, c Design diagrams of (a) bending moment and shear in the composite column, (b) of pressure at steel-concrete interface and resulting bending moment and shear in the steel and (c) in the concrete section over the connection length L_e due to $M_{Ed,a}$ corresponding to a triangular pressure distribution at the steel-concrete interface on the outstand of the embedded profile

side of the column section, as shown in Figure 5.16. Those stirrups should be present over the whole length L_a and extend with the same density over a length $L_a/6$ further to both ends of the vertical steel profile.

5.3.2 Experimental evidence

In order to validate the design method suggested hereabove, four specimens have been tested at INSA Rennes in the research project SMARTCOCO

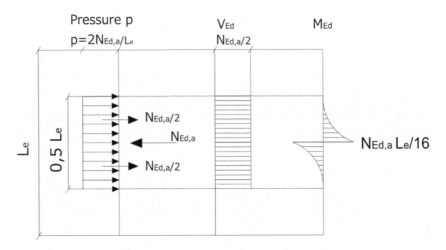

Figure 5.14 Design diagrams of pressure at the steel-concrete interface and of bending moment and shear in the steel and in the concrete section over the connection length due to $N_{Ed,a}$

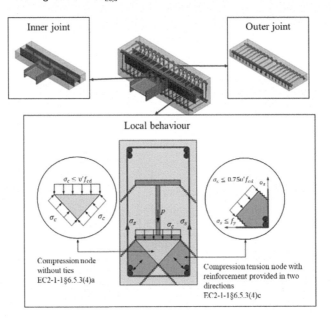

Figure 5.15 Inner and outer components of the joint and "suspension" system in the transfer zones

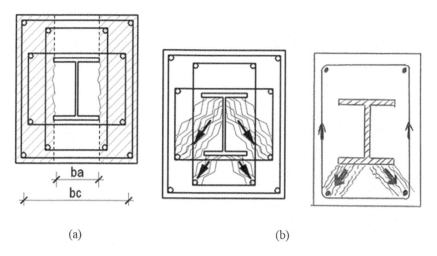

Figure 5.16 a, b (a) Effective concrete section in the connection zone for the check of the shear resistance; (b) stirrups bents facing compression struts

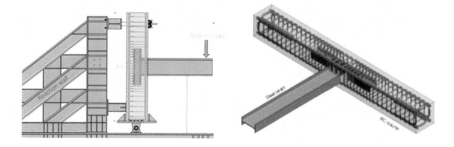

Figure 5.17 Test setup and test specimens for connections with vertical embedded profile

(Degée et al., 2017). Figure 5.17 illustrates the experimental setup and the tested specimens. The parameters varied in the four tests were the concrete class (C40 or C60) and the length of the vertical embedded profiles (1m or 1.5m).

The test specimens were instrumented with an important number of strain gauges located on the profiles and reinforcement bars in order to identify the part of the moment that is taken over by the steel profile.

The theoretical resistance of the specimen has been evaluated according to the method described hereabove and illustrated in the design example in Section 5.5.1, considering the three possible failure modes expected in the connection zone (see Figure 5.15), namely:

- The inner part of the joint (steel).
- The outer part of the joint (reinforced concrete).

Figure 5.18 Failure mode of the hybrid joints

- The transition zones (strut-and-tie mechanisms activated to transfer the load from steel to concrete).

The detailed calculations are given in Degée et al. (2017).

The failure modes are presented in Figure 5.18, showing that the behaviour is actually governed by the outer part of the joint, characterized by a shear failure of the concrete.

Table 5.1, taken from Degée et al. (2017), shows a comparison of the application of the design method to the observed maximum bending moment reached during the test. This table considers the minimum of the shear resistance of the outer and of the inner part of the joint as theoretical "web panel in shear" value, from which the corresponding bending moment is derived and compared with the maximum experimental moment. These results show a good agreement between the experimental results and the predicted values, with a range of variation of +/- 6% with a design based on the measured material properties.

Regarding the embedded length, Table 5.2 compares the minimum required length according to the suggested design method with one of the specimens, showing that the actual length was more than sufficient to

Table 51 Resistance of the node – comparison between design and experiments.

	Web panel in shear			Web in tension			Web in compression		
	$M_{wp,Rd}^{exp}$	$M_{wp,Rd}^{design}$	Error	$M_{t,Rd}^{exp}$	$M_{t,Rd}^{design}$	Error	$M_{c,Rd}^{exp}$	$M_{c,Rd}^{design}$	Error
	[kNm]	[kNm]	[%]	[kNm]	[kNm]	[%]	[kNm]	[kNm]	[%]
HJS1	1436.33	1408.68	−1.93	1076.78	1027.38	−4.59	1403.52	1415.75	0.87
HJS2	1493.79	1513.94	1.35	1097.01	1149.95	4.83	1647.81	1692.21	2.69
HJS3	1511.98	1429.77	−5.44	1136.96	1051.93	−7.48	1481.38	1471.13	−0.69
HJS4	1420.69	1516.78	6.76	1162.29	1153.26	−0.78	1783.30	1699.66	−4.69

Table 5.2 Required embedded length.

	HJS1	HJS2	HJS3	HJS4
L_{eff} [mm]	66.23	47.43	61.36	47.07
$L_{e,min}$ [mm]	610.46	572.86	600.71	572.14
$L_{e,real}$ [mm]	1000	1000	1500	1500

guarantee a proper load transfer. This is confirmed by the experimental observations, showing no evidence of damage in the transfer zone as well as no significant influence of the actual embedded length on the observed failure mode, the transfer being fully achieved in the composite part close to the node, without activating the end of the embedded profile.

5.4 TRANSITION BETWEEN COMPOSITE AND REINFORCED CONCRETE COLUMNS

The transition zone in columns that are partly reinforced concrete and partly composite, as illustrated in Figure 5.3, is also a configuration that is not explicitly covered by any available design standard or other technical references. Such situations have however been designed in the practice, as illustrated in Chapter 1. Although the origin of the action effect and the resulting distribution of pressure at the steel-concrete interface may be significantly different from what occurs in the situation covered by Section 5.3, the behaviour at a local level in the critical zone is actually very similar regarding the bending and shear behaviour. Therefore, in a way similar to what is done in Section 5.3 for framing beams connected to a reinforced concrete component through a vertical embedded piece of the profile, a design procedure is suggested by combining elements of existing standards. This procedure has been validated by experimental tests that will be summarized hereafter, followed by a detailed design example given in Section 5.5.2.

5.4.1 Design procedure

The design action effects on the composite side in column section A (Figure 5.19) are M_{Ed}, V_{Ed} and N_{Ed}. They can be subdivided into a part taken by the steel section (index a) and a part taken by reinforced concrete (index c) according to Equations (5.12) to (5.15), as formulated in Section 5.3.1.

The design should be such that $M_{Ed,a}$, $V_{Ed,a}$ and $N_{Ed,a}$ be transferred to the concrete between section A and section B over a "connection" or "embedment" length L_e (see Figure 5.19).

The embedment length L_e should in any case comply with: $L_e \geq 4\,h$ where h represents the steel section height of the profile embedded in the column for the bending axis on which $M_{Ed,a}$ is applied.

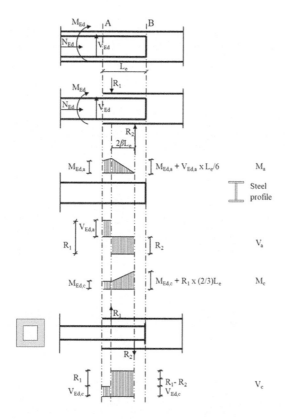

Figure 5.19 Action effects in the transition zone

The parts of M_{Ed} and V_{Ed} taken by the concrete, respectively $M_{Ed,c}$ and $V_{Ed,c}$, are directly transmitted to concrete along AB and further to section B. Only the parts of M_{Ed} and V_{Ed} taken by the steel section, respectively $M_{Ed,a}$ and $V_{Ed,a}$, require calculations for their transmission to concrete.

A realistic statically allowable solution in which equilibrium is satisfied is suggested. The steel section embedded into the concrete over the embedment length L_e (Figure 5.19) is considered a beam with simple supports at points 1 and 2 within the embedment length.

Supposing a linear distribution of stresses at the steel section-concrete interface and a distance between points 1 and 2 equal to $2/3$ L_e (common value for coupling beams, as shown in Section 5.2), the moment equilibrium at point 2 gives:

$$R_1 = \frac{\left(\dfrac{3}{2} M_{Ed,a} + \dfrac{5}{4} V_{Ed,a} \times L_e\right)}{L_e}$$

(5.16)

while moment equilibrium about point 1 gives:

$$R_2 = \frac{\left(\frac{3}{2}M_{Ed,a} + \frac{1}{4}V_{Ed,a} \times L_e\right)}{L_e} \tag{5.17}$$

The resulting diagrams of bending moment and shear in the steel section over the "connection length" L_e due to internal forces in the connection are shown in Figure 5.19. These connection effects are additional to $M_{Ed,a}$ and $V_{Ed,a}$ coming from the global analysis of the structure. This is considered in the expression given hereunder.

The checks of the steel section in connection zone AB should be made for the most unfavourable effects. These latter depend on the sign of $M_{Ed,a}$ and $V_{Ed,a}$:

Maximum design moment:

$$M_{Ed,a}or\left(M_{Ed,a} + V_{Ed} \times \frac{L_e}{6}\right) \tag{5.18}$$

Maximum design shear:

$$V_{Ed,a} \ or \ R_2 \tag{5.19}$$

In reaction, the concrete section in the connection length AB is submitted to R_1 and R_2 resulting in the diagrams of bending and shear in the concrete section shown in Figure 5.19.

These connection effects are additional to $M_{Ed,c}$ and $V_{Ed,c}$ established by the global analysis of the structure. This is also considered in the expression given hereunder.

The checks for shear of reinforced concrete in the connection zone AB should be made with a concrete section with width $b_c = b - b_a$, with b_a being the width of the steel section (see Figure 5.15b). The most unfavourable effects should be considered; they depend on the sign of $M_{Ed,c}$ and $V_{Ed,c}$:

Maximum design moment:

$$M_{Ed,c} \ or \ \left(M_{Ed,c} + R_1 \times \frac{2L_e}{3}\right) \tag{5.20}$$

Maximum design shear:

$$V_{Ed,c} \ or \ \left(V_{Ed,c} + R_1\right) \tag{5.21}$$

If stirrups are perpendicular to the element axis and if the compression struts inclination θ is taken equal to 45°, with the inner forces lever arm z assumed as $0.9\,h_{column}$ and concrete with $f_{ck} \leq 60$ MPa (meaning a strength reduction factor ν_1 for concrete cracked in shear equal to 0.6), then:

$$V_{Rd,max} = \frac{b_c \times 0.9h \times 0.6 \times f_{cd}}{2} = 0.27 b_c h f_{cd} \tag{5.22}$$

The position of the longitudinal and transverse reinforcement should be such that they support the compression struts transmitting reactions R_1 or R_2, meaning that part of or all bents should be facing the compression struts in order to constitute supporting reinforcement bringing up the reaction R_1 or R_2 to the other side of the section, as shown in Figure 5.16b. Those stirrups should be present in reaction zones R_1 and R_2, which should extend over a length $L_e/3$ at both ends of the connection zone and with the same density over a length $L_e/3$ in the reinforced concrete column.

The axial force $N_{Ed,a}$ in the steel section can be transmitted to concrete along the connection length AB by bond, by shear connectors, by end plates, by stiffeners or by a combination of those transfer means. All required information is given in Chapter 2 of this book.

Figures 5.20 and 5.21 present the possible transfer by means of end plates in the compression and tension cases respectively. The axial force $N_{Ed,a}$ is transmitted through inclined concrete compression struts to the wider concrete section (compression case) or through inclined concrete compression struts and bonds to the longitudinal bars (tension case).

If $N_{Ed,a}$ is a compression force, it should be checked that it is acceptable at the end plate:

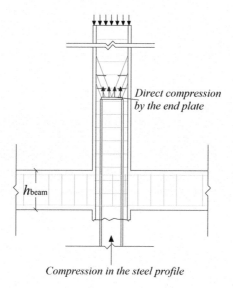

Direct compression by the end plate

Compression in the steel profile

Figure 5.20 Transmission of compression $N_{Ed,a}$ of the steel profile by means of an end plate

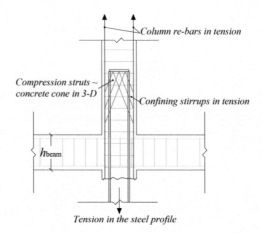

Figure 5.21 Transmission of tension $N_{Ed,a}$ of the steel profile by means of an end plate

$N_{Ed,a} \leq F_{Rdu} = A_{c0} f_{cd} (A_{c1} / A_{c0})^{0.5} \leq 3.0 f_{cd} A_{c0}$ according to Eurocode 2 Part 1 (2004) Eq. (6.63), where A_{c0} is the end plate area and A_{c1} is the concrete section area in B ($A_{c1} = b \, h$).

F_{Rdu} should in principle be reduced if the load is not uniformly distributed on the area A_{c0} (in case of a combination $M_{Ed,a}$–$N_{Ed,a}$) or if high shear forces $V_{Ed,a}$ exist. However, the connection system is such that $M_{Ed,a}$ and $V_{Ed,a}$ are not transferred to the purely concrete section B by the end plates, so that this restriction does not apply to end plate transfer of compression.

Ties should be provided to equilibrate transverse tension and/or provide confinement.

If $N_{Ed,a}$ is a tension force, it should be checked that:

- The compression force $N_{Ed,a}$ induced in compression struts at the end plate is acceptable for the compression struts.
- The pressure on the end plate is acceptable for the end plates.
- Ties are provided to equilibrate transverse tension; the sections of these ties are additional to the sections calculated for shear.
- Bond on longitudinal reinforcement provides enough resistance to transmit $N_{Ed,a}$.

The design of end plates is explained in detail in Eurocode 3 part 1-8 (2005). The verification method considers the many parameters defining the problem: the height of the column section, its width, the thickness of flanges, the presence of extension of plates outside of the steel section, etc. The design process is almost always iterative.

A simplified design consists in taking $t_p = 1.2t_f$, where t_p is the plate thickness and t_f is the thickness of the flanges.

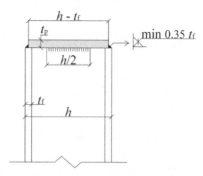

Figure 5.22 End plate of dimensions $(h - t_f)$ x b, welded to the flanges by fillet welds of length b and throat 0.35 t_f and welded to the web on half the beam depth $h/2$

The welds connecting the end plates should be designed to provide a resistance $N_{Rd,a} \geq N_{Ed,a}$.

Figures 5.22, 5.23 and 5.24 give indications of different possibilities to design the connection by making use of plates which remain within the dimensions of the steel sections; this option does facilitate the placing of the steel profile in the reinforced concrete section in comparison to wider end plates. Figures 5.22, 5.23 and 5.24 also give indicative values of the throat of fillet welds if compression/tension is the main action effect.

5.4.2 Experimental validation

In order to validate the suggested design method, eight tests have been carried out at the University of Liege in the research project SMARTCOCO (Degée et al., 2017). As illustrated in Figure 5.25, the parameters varied in the experiments are the location of the transition zone (outside – type 1 – or

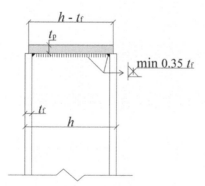

Figure 5.23 End plate of dimensions $(h - t_f)$ x b, welded to the flanges and to the web on the complete perimeter of the steel section, with weld throat equal to 0.35 t_f

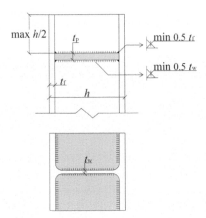

Figure 5.24 End plate placed inside the flanges. Welding on both sides of the plate connects the plate to the flanges and to the web on the complete perimeter of the steel section

inside – type 2 – the beam-column node) and the length of the transition zone (i.e. the length of the embedded profile with respect to the position of the node or the maximum moment in the column). Moreover, for each geometry, two levels of pre-compression (75 kN or 300 kN) are considered before applying the bending to failure.

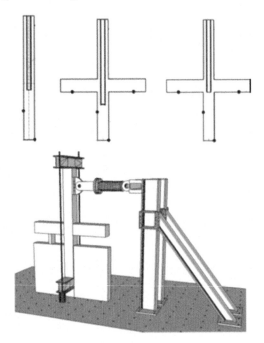

Figure 5.25 Test setup and specimens for assessing transition zones

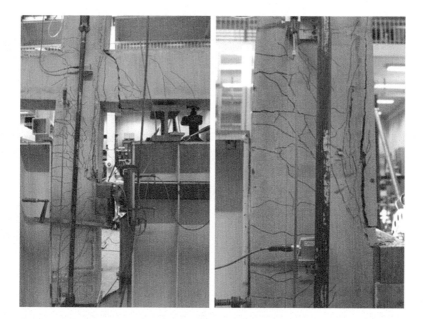

Figure 5.26 Typical failure modes of hybrid specimens

Typical failure modes are illustrated in Figure 5.26, showing that the weak point is not located in the transition zone. Table 5.3 compares the resistance obtained for the different test specimens with the predicted values resorting to the suggested design method for the transition zone. Those results show that, in all cases, the actual resistance is in the range of the predicted one with a maximum overestimation of 12% in the case of the long transition zone. This observation, coupled with the conclusions of the tests on hybrid nodes in Section 5.3 tends to suggest not to increase too much the length of the transition zones.

5.5 DESIGN EXAMPLES

5.5.1 Connection of a steel beam to a reinforced concrete column

The first worked example consists in analysing the configuration illustrated in Figure 5.2 to evaluate the maximum bending moment that can be transferred from the steel beam to the concrete column by resorting to a piece of steel profile embedded in the column and rigidly connected to the steel beam.

The assessment is carried out according to the procedure described in Section 5.3 and considers the following possible failure modes:

Table 5.3 Comparison of test results and design values for specimens for the assessment of transition zones

Specimen	Length of the transition zone [mm]	Axial compression force [kN]	Predicted shear force [kN]	Predicted bending moment [kNm]	Max. experimental shear force [kN]	Max. experimental bending moment [kNm]
1a-S-N00	320	75	102	183	200	400
1a-S-N05	320	300	102	183	204	409
1b-L-N00	635	75	304	546	208	477
1b-L-N05	635	300	340	546	216	497
2-LB-N00	635	75	280	514	278	426
2-LB-N05	635	300	280	514	310	515
3-SB-N00	320	75	317	570	318	390
3-SB-N05	320	300	317	570	308	406

1. Resistance of the steel beam.
2. Resistance of the concrete column outside the transfer zone.
3. Inner part of the composite joint in shear.
4. Outer part of the composite joint in shear.
5. Load transfer from embedded steel to concrete by local strut-and-tie mechanisms.
6. Longitudinal shear transfer from embedded steel to concrete through bond or shear studs.

The material and geometric properties considered in this example are as follows:

Material properties

Concrete class C60:	f_{ck} = 60 MPa;	γ_c = 1.5.
Reinforcing steel B500S:	f_{ysk} = 500 MPa;	γ_S = 1.15.
Structural steel S355:	f_{yk} = 355 MPa;	γ_S = 1.00.
Shear studs Nelson:	f_u = 450 MPa;	γ_v = 1.25.

Reinforced concrete column

Cross-section dimensions: b = 400 mm; h = 500 mm.
Longitudinal reinforcement: 8 ϕ25 mm.
Transverse reinforcement: stirrups ϕ10 mm – spacing of 80 mm in the joint zone and 100 mm outside the joint.
Distance between points of zero-moment according to Figure 5.13a is equal to 3.6 m.

Encased steel profile

Length of the embedded piece = 1.3 m. Steel section HEM200.
b = 206 mm; h = 220 mm; t_f = 25 mm; t_w = 15 mm; A = 13130 mm².

Steel beam

Steel section IPE300.
b = 150 mm; h = 300 mm; t_f = 10.7 mm; t_w = 7.1 mm; $W_{pl, y}$ = 628000 mm³.
The shear span ratio av of the framing in steel beam is assumed to be equal to 2.0 (i.e. $M_{Ed, max}/V_{Ed}$ = 2.0).

Nelson studs
Six studs ϕ16mm located as in Figure 5.27.

Figure 5.27 Steel part of the connection zone, including the positioning of the shear studs welded at both sides of the embedded piece of profile

1. *Resistance of the steel beam*

$$M_{Rd} = \frac{W_{pl}f_y}{\gamma_{M0}}$$ (5.23)

$W_{pl} = 628000$ mm^3; $f_y = 355$ MPa;

$$M_{Rd} = \frac{628000 \times 355}{1.0} \times 10^{-6} = 222.9\,kNm$$

2. *Shear resistance of the reinforced column outside the joint region*

$$V_{Rd} = \min\left(V_{Rd,s}; V_{Rd,max}\right)$$ (5.24)

$$V_{Rd,s} = \frac{A_{sw}}{s}\,zf_{y,wd}\cot\theta$$ (5.25)

$$V_{Rd,max} = \alpha_{cw}b_w z\nu_1 f_{cd}\frac{\cot\theta}{\cot\theta^2 + 1}$$ (5.26)

With
   spacing of stirrup outside the joint region: $s = 100$mm.
   Total area of shear reinforcement: $A_{sw} = 157.1$ (mm^2).
   Coefficient taking account of the state of the stress in the compression chord (acc. to EN 1992-1-1 for non-prestressed structures): $\alpha_{cm} = 1$.
   Strength reduction factor for concrete cracked in shear: $\nu_1 = 0.6$.
   Breadth of RC components outside of joint region: $b_w = b_c = 400$mm.
   Inner lever arm: $z = 402.75$ mm.
   Yielding

$$V_{Rd,s} = \frac{157.1}{100} \times 402.75 \times 435 \times 1 = 275233\,N$$

$$V_{Rd,max} = 1 \times 400 \times 402.75 \times 0.6 \times 40 \times \frac{1}{1+1} = 1933200\,N$$

$$V_{Rd} = \min(275233, 1933200) = 275.23\,kN$$

Based on the moment and shear diagrams given in Figures 5.13a and c, this corresponds to a moment in the framing steel beam equal to

$$M_{Rd} = V_{Rd}L = 275.233 \times 3.6 = 990.84\,kNm$$

3. Shear resistance of the inner part of the composite joint

The separation of the inner and outer parts of the joint is illustrated in Figure 5.15. The inner joint includes therefore the embedded steel profile and two layers of concrete on top and bottom of the flanges and having the same width as the profile. It corresponds to a typical connection type as considered by Eurocode 4 in its figure 8.1(2). Five different possible failure mechanisms are evaluated to assess the resistance of the inner part of the joint.

1) Column web panel in shear:

The column web panel in shear includes two components, the steel shear strength and the concrete shear strength:

$$V_{Rd} = V_{wp,Rd} + V_{wp,c,Rd} \tag{5.27}$$

Steel shear strength (acc. EN 1993-1-8 §6.2.6.1):

$$V_{wp,Rd} = \frac{0.9 f_{y,wc} A_{vc}}{\sqrt{3}\gamma_{M0}} \tag{5.28}$$

With

$$\gamma_{M0} = 1$$

Shear area of the column

$$A_{vc} = A - 2bt_f + (t_w + 2r_c)t_f$$

$$A_{vc} \geq \eta t_w h_w$$

$A = 13130$ mm^2; $b = 206$ mm; $t_f = 25$ mm; $t_w = 15$ mm; $r_c = 18$ mm; $h_w = 170$ mm; $\eta = 1.2$

$$A_{vc} = 13130 - 2 \times 206 \times 25 + (15 + 2 \times 18) \times 25 = 4105 \, mm^2$$

$$A_{vc} = 4105 > 1.2 \times 15 \times 170 = 3060$$

Resulting in a steel shear resistance component of

$$V_{wp,Rd} = \frac{0.9 \times 355 \times 4105}{\sqrt{3} \times 1} = 757222 \, N$$

Concrete shear strength (acc. EN 1994-1-1 §8.4.4.1):

$$V_{wp,c,Rd} = 0.85 v \, A_c \, f_{cd} \sin\theta \tag{5.29}$$

The coefficient taking account of long-term effects on the compressive strength and of unfavourable effects resulting from the way the load is applied, taken as $\alpha_{cc} = 1$.

The design value of cylinder compressive strength of concrete is then

$$f_{cd} = \frac{\alpha_{cc} f_{ck}}{\gamma_c} = \frac{1 \times 60}{1.5} = 40 \, MPa$$

- Lever arm for welded connection:

$$z = h_b - t_{fb} = 300 - 10.7 = 289.3 \, mm$$

- Calculation of θ:

$$\theta = arctan\left(\frac{h_s - t_{fs}}{z}\right) = arctan\left(\frac{220 - 25}{289.3}\right) = 30.44°$$

And $\sin\theta = 0.507$; $\cos\theta = 0.862$

- Cross-sectional area of concrete:

$$A_c = 0.8(b_c - t_w)(h - 2t_f)\cos\theta$$

$b_c = 206$ mm; $t_f = 25$ mm; $t_w = 15$ mm; $h = 220$ mm; $\cos\theta = 0.862$

$$A_c = 0.8 \times (206 - 15) \times (220 - 2 \times 25) \times 0.862 = 22396 \, mm^2$$

- Reduction factor for the effect of longitudinal compression in column:

$$v = 0.55\left(1 + 2\frac{N_{Ed}}{N_{pl,Rd}}\right) \le 1.1$$

With $N_{Ed} = 0 \Rightarrow \nu = 0.55$

The resulting contribution of the concrete to the shear strength is

$$V_{wp,c,Rd} = 0.85 \times 0.55 \times 22396 \times 40 \times 0.507 = 212334\,N$$

And the total column web panel resistance is

$$V_{Rd} = 757222 + 212334 = 969556\,N = 969.56\,kN$$

The relation between shear and bending moment in the connection zone is illustrated in Figure 5.15a, leading to a corresponding maximum bending moment in the framing steel beam of

$$M_{Ed,max} = \frac{V_{Rd}}{\left(\dfrac{1}{L} - \dfrac{1}{z}\right)} = \frac{969.556}{\left(\dfrac{1}{3600} - \dfrac{1}{289.3}\right)} = 305003\,kNmm = 305.00\,kNm$$

2) <u>Column web panel in transverse compression</u>

The column web panel in transverse compression includes two components, the steel and the concrete transverse compression strength:

$$F_{c,Rd} = F_{c,wc,Rd} + F_{c,wc,c,Rd} \tag{5.30}$$

Steel transverse compression strength (acc. EN 1993-1-8 §6.2.6.2):

$$F_{c,wc,Rd} = \frac{\omega k_{wc} \rho b_{eff,c,wc} t_{wc} f_{y,wc}}{\gamma_{M0}} \tag{5.31}$$

With

- Effective width of column web in compression:

$$b_{eff,c,wc} = t_{fb} + 2\sqrt{2}a_b + 5(t_{fc} + s) \tag{5.32}$$

$t_{fb} = 10.7$ mm; $a_b = 10$ mm; $t_{fc} = 25$ mm; $s = r_c = 18$ mm (rolled section)

$$b_{eff,c,wc} = 10.7 + 2\sqrt{2} \times 10 + 5 \times (25 + 18) = 254\,mm$$

- Reduction factor for the possible effect of interaction with the shear in the column (acc. EN 1994-1-1 § 8.4.4.2):

$$\omega = \omega_1 = \frac{1}{\sqrt{1+1.3\left(b_{eff,c,wc}\dfrac{t_{wc}}{A_{vc}}\right)}}$$ (5.33)

t_{wc} = 15 mm; $b_{eff,c,wc}$ = 254 mm; A_{vc} = 4105 mm²

$$\omega = \frac{1}{\sqrt{1+1.3\times\left(254\times\dfrac{15}{4105}\right)}} = 0.673$$

- Reduction factor for the effects of axial force and bending moment in the column (acc. EN 1993-1-8 § 6.2.6.2):

$k_{wc} = 1$

- Plate slenderness:

$$\bar{\lambda}_p = 0.932\sqrt{\frac{b_{eff,c,wc}d_{wc}f_{y,wc}}{Et_{wc}^2}}$$ (5.34)

$$d_{wc} = h_c - 2\left(t_{fc} + r_c\right) = 220 - 2(25+18) = 134\,mm$$

$b_{eff,c,wc}$ = 254 mm; t_{wc} = 15 mm²; E = 210000 MPa; $f_{y,wc}$ = 355 MPa

$$\bar{\lambda}_p = 0.932\times\sqrt{\frac{254\times134\times355}{210000\times15^2}} = 0.471$$

$\bar{\lambda}_p$ being smaller than 0.72, the reduction factor for plate buckling ρ is equal to 1.0

The resulting steel component of the resistance to transverse compression is equal to

$$F_{c,wc,Rd} = \frac{0.673\times1\times1\times254\times15\times355}{1.0} = 910488\,N$$

Concrete transverse compression strength (acc. EN 1994-1-1 §8.4.4.2):

$$F_{c,wc,c,Rd} = 0.85k_{wc}t_{eff,c}\left(b_c - t_w\right)f_{cd}$$ (5.35)

With

- Effective length of concrete in compression:

$$t_{eff,c} = b_{eff,c,wc} = t_{fb} + 2\sqrt{2}a_b + 5\left(t_{fc} + s\right) = 254\,mm$$

- Reduction factor for the effects of axial force and bending moment in the column:

$$k_{wc,c} = 1.3 + 3.3 \frac{\sigma_{com,c,Ed}}{f_{cd}} \le 2.0$$

$N_{Ed} = 0 \Rightarrow \sigma_{com,c,Ed} = 0 \Rightarrow k_{wc,c} = 1.3$

The resulting concrete component of the resistance to transverse compression is equal to:

$$F_{c,wc,c,Rd} = 0.85 \times 1.3 \times 254 \times (206 - 15) \times 40 = 2144186\,N$$

The total resulting resistance to transverse compression is then equal to:

$$F_{c,Rd} = 910488 + 2144186 = 3054674\,N = 3054.67\,kN$$

Given the force distribution illustrated in Figure 5.13a, the corresponding bending moment in the framing steel beam is obtained as

$$M_{Ed,max} = F_{c,Rd}z = 3054.674 \times 289.3 = 883717\,kNmm = 883.72\,kNm$$

3) Column web panel in transverse tension

The web panel resistance in transverse tension includes only a steel component, calculated according to EN 1993-1-8 §6.2.6.6:

$$F_{t,wc,Rd} = \frac{\omega b_{eff,t,wc} t_{wc} f_{y,wc}}{\gamma_{M0}} \tag{5.36}$$

With the effective width of the column web in tension equal to

$$b_{eff,t,wc} = t_{fb} + 2\sqrt{2}a_b + 5(t_{fc} + s) = b_{eff,c,wc} = 254\,mm$$

The resulting resistance to transverse tension is equal to

$$F_{t,wc,Rd} = \frac{0.673 \times 254 \times 15 \times 355}{1} = 910488\,N = 910.49\,kN$$

With a corresponding bending moment in the steel beam equal to

$$M_{Ed,max} = F_{t,wc,Rd}z = 910.488 \times 289.3 = 263404\,kNmm = 263.404\,kNm$$

4) Column flange in bending

The column flange resistance in bending includes only a steel component, calculated according to EN 1993-1-8 §4.10 and 6.2.6.4.

$$F_{fc,Rd} = \frac{b_{eff,b,wc}t_{fb}f_{y,fb}}{\gamma_{M0}} \tag{5.37}$$

With the effective breadth of steel column:

$$b_{eff,b,wc} = t_{wc} + 2s + 7kt_{fc} \tag{5.38}$$

t_{fc} = 25 mm; $t_p = t_{fb}$ = 10.7 mm; $f_{y,f}$ = 355 MPa; $f_{y,p}$ = 355 MPa

$$k = \frac{t_{fc}}{t_p} \frac{f_{y,f}}{f_{y,p}} = 2.34 > 1 \Rightarrow k = 1$$

$$b_{eff,b,wc} = 15 + 2 \times 18 + 7 \times 1 \times 25 = 226 \, mm$$

The resulting resistance of the flange in bending is equal to

$$F_{fc,Rd} = \frac{226 \times 10.7 \times 355}{1} = 858461N = 858.461kN$$

With a corresponding bending moment in the steel beam equal to

$$M_{Ed,max} = F_{fc,Rd}z = 858.461 \times 0.2893 = 248.35kNm$$

5) Bending resistance of the embedded piece of profile

The plastic bending resistance of the embedded piece of HE200M is obtained as

$$M_{pl,Rd} = \frac{W_{pl}f_y}{\gamma_{Mo}} = \frac{1135000 \times 355}{1.0} 10^{-6} = 402.93kNm$$

Based on Figure 5.13b illustrating the relation between the moment in the embedded piece of profile and in the main steel beam, it corresponds to a maximum moment in the main steel beam given by

$$M_{Ed,max} = \frac{2L_a(4L_a + 3z)M_{pl,Rd}}{(L_e - z)^2} = \frac{2 \times 505.3 \times (4 \times 505.3 + 3 \times 298.3) \times 351.913}{(1300 - 289.7)^2}$$

$$= 1162.4kNm$$

The resistances obtained for the five evaluated failure modes are summarized in Table 5.4.

Table 5.4 Resistance of the inner part of the composite joint

Web panel in shear		Web panel in transverse compression		Web panel in transverse tension		Flange in bending		Cross-section elastic bending resistance	
$V_{wp,Rd}$	$M_{Ed,max}$	$F_{c,w,Rd}$	$M_{Ed,max}$	$F_{t,w,Rd}$	$M_{Ed,max}$	$F_{fc,Rd}$	$M_{Ed,max}$	$M_{pl,Rd}$	$M_{Ed,max}$
kN	kNm	kN	kNm	kN	kNm	kN	kNm	kNm	kNm
969.6	305.0	3054.7	883.7	910.5	263.4	858.5	248.4	402.9	1162.4

4. *Shear resistance of the outer part of the composite joint*

According to Figure 5.15, the outer part of the composite joint consists of a conventional reinforced concrete section with a breadth reduced to the part of the cross-section outside the steel profile, i.e. $b_{eff} = b_c - b_p = 400 - 206 = 194$ mm. Its resistance is calculated according to the classical reinforced concrete procedure in EN 1992-1-1:

$$V_{Rd} = \min\left(V_{Rd,s}; V_{Rd,max}\right) \tag{5.39}$$

with

$$V_{Rd,s} = \frac{A_{sw}}{s} z f_{y,wd} \cot\theta \tag{5.40}$$

and

$$V_{Rd,max} = \alpha_{cw} b_w z v_1 f_{cd} \frac{\cot\theta}{\cot\theta^2 + 1} \tag{5.41}$$

With

- Total area of shear reinforcement:

$d_w = 10$ mm; $n_w = 2$ (= number of stirrup branches)

$$A_{sw} = n_w \pi \frac{d_w^2}{4} = 2 \times \pi \times \frac{10^2}{4} = 157.1\,mm^2$$

- Inner lever arm:

$a = 30$ mm (concrete cover); $d_s = 25$ mm (diameter of longitudinal reinforcements)

$$z = 0.9\left(h - a - d_w - \frac{d_s}{2}\right) = 0.9 \times \left(500 - 30 - 10 - \frac{25}{2}\right) = 402.75\,mm$$

- $\cot\theta = 1$ is taken in the shear strength calculation (chord angle equal to 45°)
- Strength reduction factor for concrete cracked in shear $v_1 = 0.6$
- Design yield strength of the shear reinforcement:

$$f_{yd} = \frac{f_{yk}}{\gamma_s} = \frac{500}{1.15} = 435\,MPa$$

- Coefficient taking account of the state of the stress in the compression chord α_{cm} equal to 1.0
- Total breadth of the outer part of the joint:

$$b_w = b_c - b = 400 - 206 = 194\,mm$$

This results in a resistance of the outer part of the joint given by

$$V_{Rd,s} = \frac{157.1}{80} \times 402.75 \times 435 \times 1 = 344042\,N$$

$$V_{Rd,max} = 1 \times 194 \times 402.75 \times 0.6 \times 40 \times \frac{1}{1+1} = 942840\,N$$

$$V_{Rd} = \min(344042, 942840) = 344042\,N = 344.042\,kN$$

Based on Figure 5.13c, illustrating the relation between the moment in the concrete part of the hybrid system and in the main steel beam, it corresponds to a maximum moment in the main steel beam given by

$$M_{Ed,max} = \frac{V_{Rd}}{\left(\dfrac{1}{L} - \dfrac{3}{4L_a + 3z}\right)} = \frac{344.042}{\left(\dfrac{1}{3.6} - \dfrac{3}{4 \times 0.5053 + 3 \times 0.2893}\right)} = 452.4\,kNm$$

5. *Local strut-and-tie mechanism for the load transfer from the embedded piece of profile to the surrounding reinforced concrete*

According to Figure 5.14 illustrating the load transfer from the profile to the stirrups through compression struts, two zones should be verified, i.e. the compression node consisting only of concrete (Figure 5.15 – Local behaviour – Left) and the compression corners consisting of concrete anchored by two ties (Figure 5.15 – Local behaviour – Right).

The different conditions to be fulfilled are the following:

- For the compression node without ties, acc. to EN 1992-1-1 § 6.5.4 (4a):

$$\sigma_c \le \sigma_{Rd,max} = k_1 v' f_{cd} = v' f_{cd} \tag{5.42}$$

- For the compression corners with ties (concrete part), acc. to EN 1992-1-1 § 6.5.4 (4c):

$$\sigma_c \le \sigma_{Rd,max} = k_3 v' f_{cd} = 0.75 v' f_{cd} \tag{5.43}$$

- For the compression corners with ties (steel part):

$$\sigma_s \leq f_y \tag{5.44}$$

The equilibrium of the node yields:

$$\sigma_c \, 2b = \sigma_s \frac{A_{sw}}{s} \leq f_y \frac{A_{sw}}{s} \tag{5.45}$$

Where s is the distance between stirrups in the joint zone

The maximum allowable stress in concrete can then be obtained as

$$\sigma_{c,max} = \min\left(v'f_{cd}; 0.75v'f_{cd}; f_y \frac{A_{sw}}{2bs} \right) \tag{5.46}$$

Based on Figure 5.13b illustrating the distribution of the contact pressure p between the flanges of the profile and the concrete, and Figure 5.15 showing the relation between the contact pressure and the local stresses in the concrete $\sigma_c = p \, / \, (2 \, b_p)$ (assuming that both the top and bottom flanges are activated), the value of the bending moment in the beam corresponding to reaching the maximum allowable stress in the nodal zone can be obtained by:

$$M_{Ed,max} = \frac{L_a\left(4L_a + 3z\right)}{6} p_{Ed,max} = \frac{L_a\left(4L_a + 3z\right)}{6} 2b \min\left(v'f_{cd}; 0.75v'f_{cd}; f_y \frac{A_{sw}}{2bs} \right)$$

with

$$v' = 1 - \frac{f_{ck}}{250} = 1 - \frac{60}{250} = 0.76$$

$b = 206$ mm; $L_e = 1300$ mm; $f_{cd} = 40$ MPa; $f_y = 500$ MPa; $A_{sw} = 157.1$ mm^2; $s = 80$ mm

This leads to a value of $M_{Ed,max}$ equal to 238.9 kNm.

6. Strength of connection between encased profile and concrete.

The shear force coming from the framing steel beam is transferred as a normal force in the concrete column by the activation of two possible mechanisms, i.e. (i) the activation of the bond resistance at the steel-concrete interface between the concrete column and embedded piece of profile or (ii) the activation of the shear resistance of the six shear studs welded on the profile as given in Figure 5.27. It is assumed that the mechanism providing the larger resistance from those two is the one effectively activated.

$$V_{Rd,max} = \max\left(V_{Rd,bond}; V_{Rd,studs} \right) \tag{5.47}$$

(i) Load transfer through steel-concrete bond strength

$$V_{Rd,bond} = L_e \chi_{profile} \tau_{Rd} \tag{5.48}$$

With

Design shear strength (according to EN 1994-1-1 Table 6.6 for completely concrete encased steel sections) $\tau_{Rd} = 0.3$

Perimeter of steel profile section:

$$\begin{aligned}\chi_{profile} &= 2b + 4t_f + 2\pi r + 2(h - 2t_f - 2r) + 2(b - t_w - 2r)\\ &= 2 \times 206 + 4 \times 25 + 2\pi \times 18 + 2\,(220 - 2 \times 25 - 2 \times 18) + 2\,(206\\ &\quad - 15 - 2 \times 18)\end{aligned}$$

$\chi_{profile} = 1203$ mm

Resulting in a design longitudinal shear resistance

$$V_{Rd,bond} = 1300 \times 1203 \times 0.3 = 469208\,N = 469.2\,kN$$

(ii) Load transfer through shear studs

$$V_{Rd,studs} = nP_{Rd} \tag{5.49}$$

with

$$P_{Rd} = \frac{0.8 f_u \pi d^2}{4\gamma_v}$$

$f_u = 450$ MPa; $d = 16$mm; $\gamma_v = 1.25$

$$P_{Rd} = \frac{0.8 \times 450 \times \pi \times 16^2}{4 \times 1.25} = 57906\,N$$

Resulting in a total resistance for the six shear studs equal to

$$V_{Rd} = 6 \times 57906 = 347435\,N = 347.4\,kN$$

The shear force from the beam is therefore assumed to be transferred by bond with

$$V_{Rd,max} = max\,(469.2;\ 347.4) = 347.4\,kN$$

The corresponding bending moment in the steel beam is then derived based on the shear span ratio of the beam ($a_v = 2.0$ m) as

$$M_{Rd,max} = V_{Rd,max} a_v = 347.4 \times 2.0 = 694.8\,kNm$$

Table 5.5 Summary of the resistance corresponding to the different evaluated failure modes

Steel beam	RC outside joint region	Inner part of the joint	Outer part of the joint	Local strut-and-tie transfer mechanism	Shear transfer from steel to concrete
kNm	kNm	kNm	kNm	kNm	kNm
222.9	990.8	248.4	452.4	238.9	694.8

According to the summary given in Table 5.5, the bending moment resistance of the studied hybrid joint is $M_{Rd,max} = 222.9$ kNm, corresponding to the bending resistance of the framing steel beam.

5.5.2 Transition zone from a composite to a reinforced concrete column

This example illustrates the procedure described in Section 5.4 to design the transition zone in a column that is partly composite and partly made of reinforced concrete, as illustrated in Figure 5.3. Two cases will be considered, namely (i) a case where the transition occurs outside a structural node and (ii) a case where the transition occurs at the level of a crossing beam. Figures 5.28 and 5.29 illustrate the chosen global configurations respectively for the first (cantilever column) and second (with crossing beams) cases, with the corresponding global bending moment diagrams.

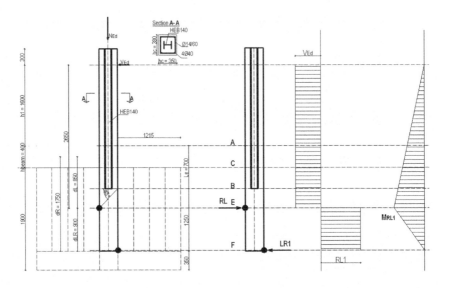

Figure 5.28 First configuration

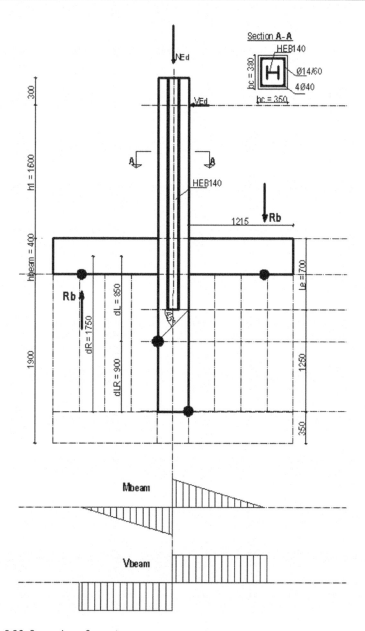

Figure 5.29 Second configuration

In both cases, the transition zone is assumed from level A to level B. This means that the verification above level A should be carried out according to a classical steel-concrete component procedure (for instance Eurocode 4 part 1-1) and, below level B as well as in the beams of configuration 2, according to a classical reinforced concrete procedure (for instance

Eurocode 2 part 1-1). Such verifications are not considered in the example, to strictly focus on zone AB.

It must be noted that the choice of the length considered as transition zone (i.e. the position of point A) remains rather arbitrary but should neither be too long (as demonstrated by the experimental results summarized in Sections 5.3 and 5.4) nor shorter than 1.5 times the height of the profile cross-section, similar to what is suggested for profile framing inside a concrete wall as referred to in Section 5.2. Typically, it is recommended to select a length in the range of four to eight times the height of the profile, as a shorter length is likely to lead to very high-stress concentrations. The transition zone should be provided with a higher density of transverse reinforcements than the regular zones of the column. In the current examples, the transition length is chosen as equal to 700 mm, which corresponds to five times the height of the profile.

Material properties

Concrete C40:	$f_{ck} = 40$ MPa
Reinforcements B500S:	$f_{sk} = 500$ MPa
Structural steel S460:	$f_{yk} = 460$ MPa

For the sake of simplicity, all partial material factors are considered equal to 1.0 in this example.

Geometric and mechanical properties
The cross-section of the composite column is given in Figure 5.30.

- Concrete cross-section $B \times H = 330 \times 350$ mm.
- Fully embedded steel profile HE140B.
- Longitudinal reinforcements $4\phi40$ mm.
- Transverse reinforcement in the transition zone: stirrups $\phi14$ mm with a spacing s of 60 mm.

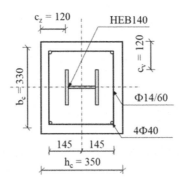

Figure 5.30 Cross-section of the composite part

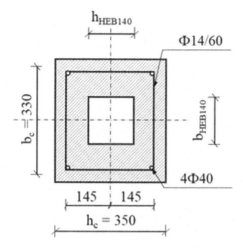

Figure 5.31 Reduced concrete cross-section

The properties of the embedded steel profile are:

$M_{pl,Rd,a}$ = 112.88 kNm

$V_{pl,Rd,a}$ = 347.38 kN

In view of further verifications, it is also relevant to evaluate the properties of the reduced concrete section in the transition zone, obtained by removing the central part occupied by the steel profile (see Figure 5.31) and evaluating its properties as a usual reinforced concrete section. The resulting properties are:

$M_{Rd,c1}$ = 375 kNm

$V_{Rd,c1}$ = 808.2 kN

Finally, the properties of the full composite cross-section can also be derived. The bending resistance of the composite cross-section is evaluated by resorting to a suitable software tool as $M_{pl,Rd}$ = 570 kNm, while the shear resistance is assumed equal to the sum of the shear resistance of the steel profile and of the reduced concrete section, equal to 1155.6 kN.

1. *Control of the transition zone*

First of all, the design action effects in the composite column section at point A are subdivided into a part taken by the steel profile and a part taken by the reinforced concrete.

$$M_{Ed} = M_{Ed,a} + M_{Ed,c} \tag{5.50a}$$

$$V_{Ed} = V_{Ed,a} + V_{Ed,c} \tag{5.50b}$$

$$N_{Ed} = N_{Ed,a} + N_{Ed,c} \qquad\qquad (5.50c)$$

According to EN 1994-1-1, the subdivision is made in proportion to the respective bending resistance.

$$M_{Ed,a} = M_{Ed} \cdot \frac{M_{pl,Rd,a}}{M_{pl,Rd}} \qquad\qquad (5.51a)$$

$$V_{Ed,a} = V_{Ed} \cdot \frac{M_{pl,Rd,a}}{M_{pl,Rd}} \qquad\qquad (5.51b)$$

where:

$$M_{pl,Rd,a} = 112.88\,kNm;$$

$$M_{pl,Rd} = 570\,kNm$$

As a result,

$$M_{Ed,a} = 0.2 \cdot M_{Ed} \quad V_{Ed,a} = 0.2 \cdot V_{Ed}$$

$$M_{Ed,c} = 0.8 \cdot M_{Ed} \quad V_{Ed,c} = 0.8 \cdot V_{Ed}$$

$M_{Ed,c}$ and $V_{Ed,c}$ are already taken by the concrete part and will remain in the concrete part until point C. The design procedure consists then in identifying a suitable mechanism allowing the transfer of $M_{Ed,a}$ and $V_{Ed,a}$ from the steel to the concrete over the transition length L_e. Such a mechanism is illustrated in Figure 5.32 and activates the transfer by creating a contact pressure at the steel-concrete interface, respectively p_M for the moment transfer and p_V for the shear transfer.

Transfer of $M_{Ed,a}$ – pressure p_M:

$$R_M = \frac{3 \cdot M_{Ed,a}}{2 \cdot L_e}, \quad p_M = \frac{4 \cdot R_M}{L_e} = \frac{6 \cdot M_{Ed,a}}{L_e^2} \qquad\qquad (5.52)$$

Transfer of $V_{Ed,a}$ – pressure p_V:

$$p_V = \frac{V_{Ed,a}}{L_e} \qquad\qquad (5.53)$$

Transfer of $M_{Ed,a} + V_{Ed,a}$:

$$p_{max} = p_M + p_V, \quad p_{min} = p_M - p_V \qquad\qquad (5.54a)$$

$$L_1 = \frac{L_e}{2} \cdot \frac{p_M + p_V}{p_M}, \quad L_2 = \frac{L_e}{2} \cdot \frac{p_M - p_V}{p_M} \qquad\qquad (5.54b)$$

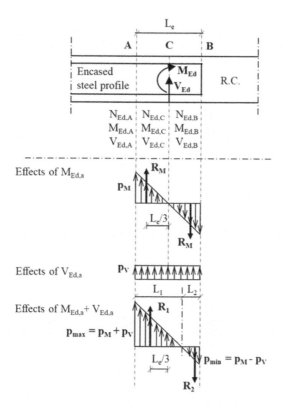

Figure 5.32 Moment and shear transfer with corresponding contact pressure distribution

$$R_1 = \frac{p_{max} \cdot L_1}{2}, \quad R_2 = \frac{p_{min} \cdot L_2}{2} \tag{5.54c}$$

The evaluated contact pressure can then be used to evaluate the total resulting bending moment and shear forces respectively in the steel profile and in the surrounding concrete for the entire transition length, as illustrated in Figures 5.33 and 5.34.

It should finally be checked that the peak values of $V_{Ed,a}$ and $M_{Ed,a}$, respectively $V_{Ed,c}$ and $M_{Ed,c}$, are lower than the corresponding resistances, $V_{Rd,a}$ and $M_{Rd,a}$, respectively, $V_{RD,c1}$ and $M_{Rd,c1}$ (resistance of the reduced concrete section).

It is relevant to note that, up to this point, the procedure is similar for the cantilever column as for the configuration with the transition zone at the level of a crossing beam.

Numerical application
Assuming for instance that the acting forces M_{Ed} and V_{Ed} are respectively equal to 350 kNm and 250 kN, with $L_e = 700$ mm, the application of the above procedure leads to:

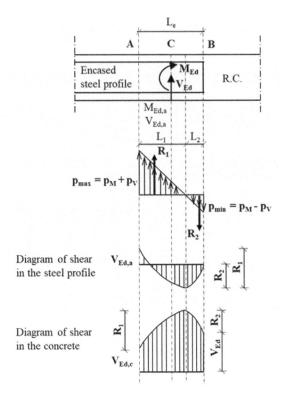

Figure 5.33 Shear force diagrams in the transfer zone AB

$M_{Ed,a}$ = 70 kNm and $V_{Ed,a}$ = 50 kN
$M_{Ed,c}$ = 280 kNm and $V_{Ed,c}$ = 200 kN
p_M = 6 x 70 / 0.7² = 857.1 kN/m
p_V = 50 / 0.7 = 71.4 kN/m
p_{max} = 857.1 + 71.4 = 928.6 kN/m
p_{min} = 857.1 − 71.4 = 785.7 kN/m
L_1 = (0.7/2) x (857.1 + 71.4) / 857.1 = 0.379 m
L_2 = (0.7/2) x (857.1 − 71.4) / 857.1 = 0.321 m
R_1 = 928.6 x 0.379 / 2 = 176.0 kN
R_2 = 785.7 x 0.321 / 2 = 126.0 kN

Verification of the steel profile:

$V_{Ed,a}$ = 50.0 kN ≤ $V_{pl,Rd,a}$ = 347.4 kN
R_2 = 126.0 kN ≤ $V_{pl,Rd,a}$ = 347.4 kN
$M_{Ed,a}$ = 70.0 kNm ≤ $M_{pl,Rd,a}$ = 112.9 kN

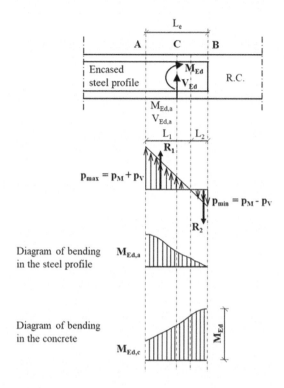

Figure 5.34 Bending moment diagrams in the transfer zone AB

Verification of the concrete in zone AB:

$$V_{Ed,c} + R_1 = V_{Ed} + R_2 = 376.0 \text{ kN} \leq V_{Rd,c1} = 808.2 \text{ kN}$$
$$M_{Ed,c} = 280.0 \text{ kNm} \leq M_{Rd,c1} = 375.0 \text{ kN}$$
$$M_{Ed} = 350.0 \text{ kNm} \leq M_{Rd,c1} = 375.0 \text{ kN}$$

2. *Resistance to the contact pressure*

Depending on the configuration around the transition zone AB, the resistance to the contact pressure is ensured either by the development of compression struts or by direct bearing. Typically, two situations can be faced, as shown in Figure 5.35.

- Case 1: the contact pressure from the steel section is resisted by concrete compression struts developing in the dashed zones of the figure.
- Case 2: in the case of beams present next to the transition zone, the contact pressure can possibly face the beam section. In this case, the pressure can be resisted by direct bearing. In this case, the contact pressure at the tip of the profile remains resisted by compression struts, as in case 1.

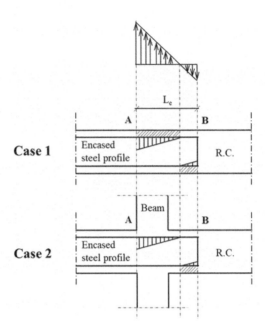

Figure 5.35 Case I and 2 – resistance mechanism to the pressure at the steel-concrete interface

According to Eurocode 2 part 1-1, the maximum stress than can be supported by a given compression strut is given by:

$$\sigma_{Rd.max} = k_3 \cdot \upsilon' \cdot f_{ck} = 0.75 \cdot 0.84 \cdot 40\,MPa = 25.2\,MPa$$

where:

$$\upsilon' = 1 - \frac{f_{ck}}{250\,MPa} = 0.84$$

With $f_{ck} = f_{cd} = 40$ MPa in this academic example.

In general, the development of the compression struts depends on the shape of the cross-section and hence on the inclination angle of the concrete struts and on the bent radius of the stirrups in the corner facing the compression chords.

Figure 5.36 shows two possibilities for a rather stocky cross-section, with strut angles equal to or smaller than 45° (note that the geometry of the current example corresponds to the left figure in Figure 5.36), while Figure 5.37 shows another possibility for a slender section with an angle greater than 45°.

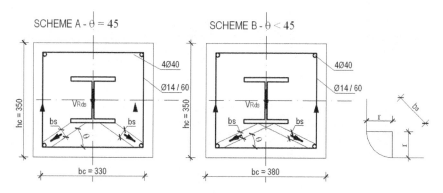

Figure 5.36 Compression struts for stocky sections – left figure corresponds to the geometry of the worked example

It is first useful to remember the minimum mandrel diameter recommended to bend the stirrups, depending on the bar diameters, namely:

- For $\phi \leq 16$ mm: $2r = 4\ \phi$;
- For $\phi > 16$ mm: $2r = 7\ \phi$;
- For schemes A and B in Figure 5.36, the width of the compression struts can then be directly related to this radius by

$$b_s = r \cdot (\cos\theta + \sin\theta) \tag{5.55}$$

leading then to total strength of the strut given by

$$F_{strut} = \sigma_{Rd,max} \cdot b_s = \sigma_{Rd,max} \cdot r \cdot (\cos\theta + \sin\theta) \tag{5.56}$$

Adequate projections provide then the resulting resistance in the vertical direction of the cross-section, which can be activated to transfer the contact pressure at the flange-concrete interface:

$$V_{Rd.s-\theta\leq45} = 2 \cdot F_{strut} \cdot \cos\left(\frac{\pi}{2} - \theta\right) = 2 \cdot F_{strut} \cdot \sin\theta \tag{5.57}$$

For scheme C, with an angle θ greater than 45°, the strut can be partly anchored in the vertical stirrups by bond over a length l_b (see Figure 5.37). In this case, the resistance to the local pressure may not be limited by the width of the compression struts but by the direct compression right under the flange of the profile, leading to a vertical resistance equal to

$$V_{Rd,s,max} = b_{HE140B} \cdot f_{cd} \tag{5.58}$$

SCHEME C - θ > 45

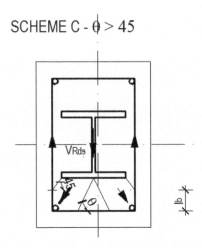

Figure 5.37 Compression struts for a slender section (θ > 45°)

similar to what has been described previously for a transition zone at the level of a crossing beam.

Numerical application

In the current example, with stirrups $\phi14$ mm, the bent radius is 28 mm.

For the cantilever column, the resistance must be ensured by the development of compression struts in the dashed zones of Figure 5.35. According to Figure 5.36 (scheme A), a strut angle of 45° can be considered, leading to,

$$b_s = r \cdot (\cos\theta + \sin\theta) = 2 \times 28 \times \cos(45°) = 39.6\,mm$$

$$F_{strut} = \sigma_{Rd,max} \cdot b_s = 25.2 \times 39.6 = 997.9\,kN\,/\,m$$

$$V_{Rd.s-\theta=45} = 2 \cdot F_{strut} \cdot \cos(45°) = 1411.3\,kN\,/\,m$$

This value can then be compared to the acting contact pressure equal to 928.6 kN/m.

For case 2, with a beam present next to the transition zone, the direct bearing resistance is given by:

$$V_{Rd,s} = b_{HE140B} \cdot f_{cd} = 140 \times 40 = 5600\,kN\,/\,m$$

Which is by far sufficient to ensure the resistance to a local contact pressure of 928.6 kN/m.

3. *Transmission of the axial force*

This example considers transmission of the compression forces in the column from steel to concrete through an end plate welded to the end of the column, similar to the configuration suggested in Figure 5.20. According to the considerations given in Figures 5.22 and 5.23, a solution with the following plate dimensions can be suitable:

- Plate width $a = b_{f\_HE140B} = 140$ mm
- Plate length $b = h_{HE140B} = 140$ mm
- Plate thickness $t_p = 10$ mm ($> 0.35\, t_{f\_HE140B}$)

The verification of the compression transmission described in Section 5.4 refers to Eurocode 2 and requires the verification of the following condition:

$$N_{Ed.a} \le F_{Rd.u} \le 3.0 \cdot A_{co} \cdot f_{cd} \tag{5.59}$$

With

$$A_{co} = b_{HE140B} \cdot b_{HE140B} = 19600\, mm^2$$

$$A_{c1} = A_{tot,c} - A_{co} = 133000\, mm^2 - 19600\, mm^2 = 113400\, mm^2$$

$$F_{Rd.u} = A_{co} \cdot f_{cd} \cdot \left(\frac{A_{c1}}{A_{co}}\right)^{0.5} = 196\, cm^2 \cdot 40\, MPa \cdot \left(\frac{1134\, cm^2}{196\, cm^2}\right)^{0.5} = 1886\, kN$$

$$3.0 \cdot A_{co} \cdot f_{cd} = 2352\, kN$$

The maximum allowable value of $N_{Ed,a}$ is thus equal to 1886 kN.
   With

$$N_{Ed.a} = N_{Ed} \cdot \frac{A_{HE140B} \cdot f_{ay}}{N_{pl.Rd,composite}} = N_{Ed} \cdot \frac{4300 \times 460}{0.85 \times 123670 \times 40 + 4300 \times 460 + 5030 \times 500}$$

$$= 0.227 \cdot N_{Ed}$$

This leads to a maximum allowable global compression load acting on the composite column equal to 8307 kN. With $N_{pl,Rd,composite}$ being equal to 8697 kN, this shows that the transition zone has only a minor impact on the compression capacity of the column.

REFERENCES

AISC 341–16. (2016) *Seismic provisions for structural steel buildings.* American Institute of Steel Construction, Chicago, IL.

Degee, H., Plumier, A., Mihaylov, B., Dragan, D., Bogdan, T., Popa, N., et al. (2017) Smart composite components – concrete structures reinforced by steel profiles "Eurocode".

Eurocode 2. (2004) *Design of concrete structures, part 1.1 – General rules for buildings*. European Committee for Standardizations, Brussels.

Eurocode 3. (2005) *Design of steel structures, part 1.1: General rules and rules for buildings*. European Committee for Standardizations, Brussels.

Eurocode 4. (2004) *Design of composite steel and concrete structures, part 1.1 – general rules for buildings*. European Committee for Standardizations, Brussels.

Eurocode 8. (2004) *Design provisions for earthquake resistance - part 1: General rules, seismic actions and rules for buildings*. European Committee for Standardization, Brussels.

Marcakis, K., and Mitchell, D. (1980) Precast concrete connections with embedded steel members. *Journal Prestressed Concrete Institute*, 25(4), 86–116.

Mattock, A. H., and Gaafar, G. H. (1982) Strength of embedded steel sections as brackets. *ACI Structural Journal*, 79(9), 83–93.

Minami, K. (1985) Beam to column stress transfer in composite structures. In *Composite and mixed construction*. Proceedings of the U.S/Japan joint Seminar, C. Roeder, ASCE, New York, 215–226.

Park, W.-S., Yun, H.-D., Hwang, S.-K., Han, B.-C., Yang, I.S. (2005) Shear strength of the connection between a steel coupling beam and a reinforced concrete shear wall in a hybrid wall system. *Journal of Construction Steel Research*, 61, 912–941.

Park, R., Priestley, M. J. N., and Gill, W. D. (1982) Ductility of square confined concrete columns. *ASCE Journal of the Structural Division*, 108(4), 929–950.

Chapter 6

Connections of reinforced concrete beams or flat slabs to steel columns using shear keys

Dan Bompa and Ahmed Elghazouli

CONTENTS

DOI: 10.1201/9781003149811-6

6.1 INTRODUCTION

Situations in which reinforced concrete floor elements need to be combined with vertical steel components can arise in multi-storey buildings, either due to loading and performance constraints or because of practical and constructional considerations. Such situations may require non-standard design approaches, beyond conventional reinforced concrete (RC) or steel-concrete composite design, and a hybrid form of detailing. Hybrid connections between steel columns and horizontal RC floors can be made by means of relatively short inserts that are fully embedded in the concrete body and welded to the steel column. The main purpose of these inserts, referred to as "shear keys", is to transfer the shear and bending moment from the horizontal RC floors to the vertical steel component.

Figure 6.1 depicts the structural configuration of interior connections between steel columns and reinforced concrete floors. The left-hand side of Figure 6.1a illustrates the in-plane symmetric configuration of a moment frame where shear keys are used to transfer the loads from the slab through beams to the steel column. The right-hand side of Figure 6.1a shows the in-plane symmetric configuration of a flat slab where loads are transferred through the

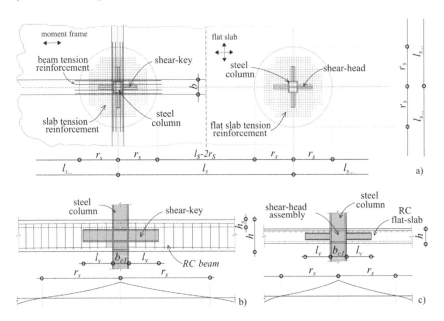

Figure 6.1 a) General scheme of hybrid connections; hybrid connections to: b) reinforced concrete beams, c) reinforced concrete flat slabs

shear-head system to the steel column. On the other hand, Figures 6.1b,c depict the side views of hybrid connections to beams and slabs. The fully integrated shear key is defined by its length l_v, its cross-section and depth in the composite cross-section. The shear key is typically located at the centre of the cross-section, to ensure even concrete cover in tension and compression. In hybrid connections to one-way components, the shear keys are welded to the flange of the column symmetrically to its centreline. In flat slabs, referred to as two-way components, cruciform shear heads have two shear keys directly welded to the column flange and two perpendicular shear keys are connected to a plate that is welded to the edges of the column flanges. For tubular columns, shear keys would be directly welded to the tube wall and could be provided with continuity plates for enhanced moment resistance.

The presence of an embedded steel element within a reinforced concrete component creates a discontinuity within two distinct zones (i.e. steel-concrete composite and non-composite reinforced concrete) and results in relatively complex behavioural characteristics. In addition to the two distinct regions of the hybrid domain, there is a third "transition" region. A number of failure modes can occur within these three regions of the hybrid connection, either in bending or in shear.

6.2 COMMON DEFINITIONS AND HYPOTHESES

As noted above, hybrid connections of beams or slabs to steel columns can be divided into several regions in which corresponding equilibrium, constitutive and compatibility laws apply. These are schematically illustrated in Figure 6.2, and are defined as follows:

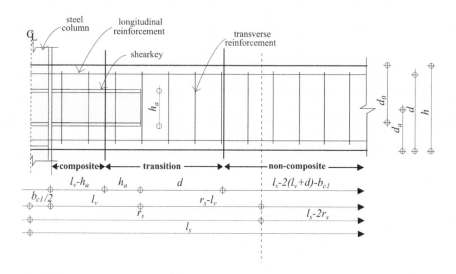

Figure 6.2 Regions of a hybrid connection to beam component

- The *joint region* contains the column web panel zone, column flanges and shear key to column-welded interface. The design of the joint region is not covered in this section and should follow the guidelines provided in Eurocode 3 (CEN, 2005) and Eurocode 4 (CEN, 2004a).
- A *composite region* that is bounded by the face of the column and a distance equal to the length of the shear key minus its depth $(l_v\text{-}h_a)$. This region may be defined largely as a composite steel-concrete component. Without the provision of shear connectors, the cross-section benefits from some degree of interaction between the steel shear key and surrounding concrete. The detailing of this region follows common rules of Eurocode 4 (CEN, 2004a) for encased profiles.
- A *transition region* between the composite and non-composite regions contains the composite-to-reinforced concrete *interface*. As this is a discontinuity within two regions with distinctive characteristics, its behaviour is more complex than that occurring in conventional composite or reinforced concrete components. This is likely to be the *critical region* both in bending and shear.
- A *non-composite region*, outside of the shear key region in which typical reinforced concrete rules apply. Longitudinal and transverse reinforcement detailing, concrete cover and other detailing requirements follow the main design rules stipulated in Eurocode 2 (CEN, 2004b).

The response of the composite region can be governed by flexure or shear with the critical cross-section typically located at the column face. Depending on the shear key geometry, reinforcement ratio and concrete section properties, the flexural response can be governed by simultaneous mechanisms: yielding in the top tensile reinforcement, yielding in the top flange of the shear key, and partial yielding in the shear key web. Due to the contribution of the shear key to the tension capacity of the cross-section along with the longitudinal reinforcement, crushing would govern in the concrete compression zone at the column face. As the steel-concrete interaction is based on the friction of the two components, the degree of shear connection is partial. Although beam shear is unlikely to govern in the composite region due to the relatively high contribution of the shear key to the cross-sectional shear resistance, verification checks for such an ultimate state should be carried out.

In both one-way and two-way components, the bottom flange of the shear key acts as support for a force-transferring strut (Figure 6.3). The inclination and orientation of this strut are influenced by the geometry of the component and that of the shear key. Relatively rigid steel shear keys act as stiff support for the force-transferring strut, whilst more flexible shear keys bend under loading with the strut support region, moving towards the column face with increased component rotations. As indicated in Figure 6.4a, when relatively low reinforcement ratios exist (i.e. ρ_l=0.3–0.5%), the critical condition is governed by flexure at the *interface*, either by yielding

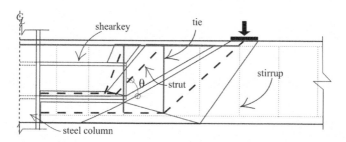

Figure 6.3 Strut-and-tie response in one-way components

of the longitudinal reinforcement or by crushing of concrete in compression. Assuming that shear reinforcement exists, in one-way components, the presence of relatively high flexural reinforcement ratios (i.e. ρ_l>1.0%), facilitates a shear critical state of stresses at the *interface* (Figure 6.4b). These combined structural parameters typically promote diagonal tension failures in which the governing shear crack passes the shear key just below the inclined strut and intersects the neighbouring shear reinforcement (Bompa and Elghazouli, 2015). For loads applied close to the shear key tip, within a distance of about the component depth, a direct strut mechanism develops, and the critical condition is controlled by strut crushing.

In an axisymmetric representation of the two-way flat slab components, as illustrated in Figure 6.5a, the zones incorporating the shear keys can be referred to as "hybrid sectors", following the orthogonal directions of the slab. Zones without shear heads (diagonals of the slab) are referred to as "reinforced concrete sectors". The hybrid sectors of flat slab connections are divided into the same number of regions as the one-way cases: a joint region, a composite region, a transition region, containing the composite-to-RC interface and a non-composite region. These are depicted in Figure 6.5b.

In hybrid flat slab connections, the composite regions are unlikely to be critical in design. Shear key systems shift the critical shear and moment regions away from the column face, exhibiting behaviour similar to that of conventional flat slabs supported by larger columns. The concrete inside the four shear keys is under a form of bi-axial or tri-axial confinement with peaks in the vicinity of the column. Moreover, the cross-sections incorporating the shear keys have inherent higher moment capacity than those outside the shear key region. Although some top flange shear key yielding is likely to occur at ultimate, in practical configurations, the weak section in bending is also at the interface. For completeness, design verifications for the composite sections should be undertaken, to ensure adequate capacity for the design requirements. In the transition region, the fundamental behaviour of hybrid connections to flat slabs is essentially the same as for one-way cases, noting that due to the flat slab geometry, the strut development is not restricted within a confined space as for beam components, but rather able to develop in a fan-shaped manner.

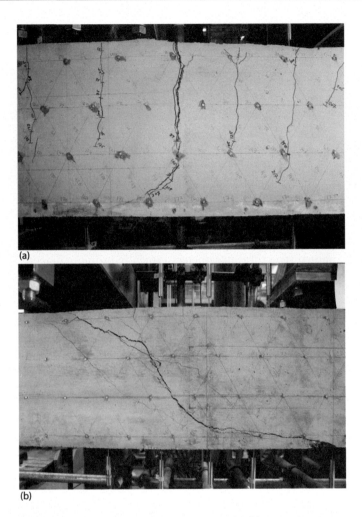

Figure 6.4 Failure kinematics of hybrid connections with: a) low reinforcement ratios, b) relatively high reinforcement ratios

In a sectional view through a hybrid sector at the shear key centre, similarly to beam components, the critical region both in flexure and shear is typically located around the *interface*, in the vicinity of the tip of the four shear keys of the cruciform shear head. For relatively high reinforcement ratios, due to the plan fan-like strut development, punching shear critical regions appear at about a half slab depth from the four shear key tips. This is illustrated in Figure 6.6. With the increase in slab rotation, the shear critical sections extend to the reinforced concrete sectors creating a critical surface, represented in the plan by a critical perimeter. When relatively low flexural reinforcement ratios exist, the response is typically governed by rebar yielding, yielding in the top flange of the shear key, potential partial

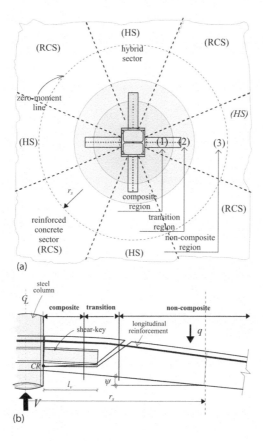

Figure 6.5 a) In-plane view of a hybrid flat slab connection, b) cross-sectional view of a hybrid flat slab connection

yielding in the shear key web, and/or crushing of the concrete in compression. Shear reinforcement can be used to enhance the punching shear strength and ductility of hybrid flat slab connections, noting that the exact contribution of the transverse reinforcement depends on the stud layout, diameter and spacing, as well as shear key system layout and geometry, slab geometry, among other factors.

6.3 CONNECTION OF RC BEAMS TO STEEL COLUMNS THROUGH FULLY EMBEDDED SHEAR KEYS

This section deals with the design of hybrid connections between one-way reinforced concrete components and steel columns by means of fully embedded shear keys, in moment frame buildings. As noted above, particular focus is given to the design of the composite and transition regions, whilst

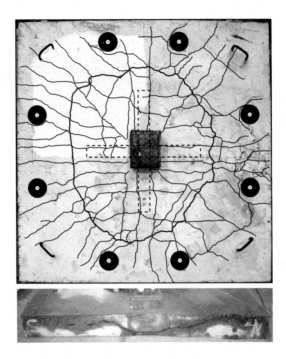

Figure 6.6 Critical sections in hybrid connections to flat slabs

the joint and non-composite regions should be designed using available procedures in respective Eurocodes. The main assumptions on the cross-sectional response are: (i) plane sections remain plane considering a linear compatibility relationship between the tension and compression fibres, (ii) concrete tensile strength is ignored in all regions, (iii) in all regions, the strain in the longitudinal reinforcement bars is the same as that in the surrounding concrete, (iv) in the composite region, the strain in the shear key is determined at reinforcement yielding by linear cross-sectional strain compatibility, (v) the composite cross-section follows the design of fully encased components, (vi) the shear key contributes to the strength and stiffness of the composite region, by means of an enhanced flexural reinforcement ratio.

6.3.1 Flexural strength

For very small shear key section sizes in relation to the embedding beam cross-section, the flexural behaviour at the column face approaches a conventional reinforced concrete component. As the shear key section size increases, the inherent bond levels between the shear key and concrete improve, resulting in a higher level of interaction between the two entities, and stronger composite action. Non-linear parametric investigations showed that the interaction increases generally with the increase in transverse reinforcement ratio

(Moharram, 2018). The latter provides confinement to the shear key, hence delaying interface slip. It is worth noting that full cross-sectional compatibility between the shear key, assuming perfect interaction between the steel insert and concrete, cannot be achieved without providing shear connectors. Numerical investigations showed that a reduction in stiffness and strength in the range of 5–20% from the full composite behaviour occurs for typical composite regions of the hybrid connection.

Due to the inherent characteristics of such connections, a full plastic distribution is unlikely to be achieved, but some yielding in the top shear key flange would develop. In such situations, the shear key design should be carried out assuming an elastic second moment of inertia. An enhanced composite action can be achieved by providing shear connectors in the form of welded studs at the compression flange of the shear key. This hypothesis requires a definition of the design longitudinal shear at the interface between the steel flange and the concrete. This is achieved by defining an effective interacting concrete section and a design compression stress in that concrete, in accordance with Eurocode 4 guidelines.

The assumptions and design steps are described in Section 6.3.4. It is worth pointing out that, although full plastic distribution can be achieved in the composite cross-sections, this would have only a localized effect and perhaps more simple detailing measures such as increasing the flexural reinforcement ratio or shear key section size would balance the need for the 5–20% supplementary moment capacity achieved without shear connectors.

The flexural strength of hybrid components should be verified for all three regions defined in Figure 6.2. In all cross-sections, the design bending moment should be less than the bending resistance of each cross-section (Figure 6.7). In the composite region and in the transition region within the length of the shear key, the bending moment demand $M_{Ed,k}$ should be smaller than the moment resistance of the composite cross-section $M_{Rd,k}$ considering elastic section properties when only partial shear connection exists (i.e. $M_{Ed,k} < M_{Rd,k}$). In the transition region outside of the shear key

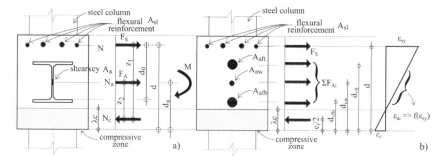

Figure 6.7 Detail of composite cross-section: a) actual configuration, b) idealized configuration

and in the non-composite region, the design moments $M_{Ed,kc}$ and $M_{Ed,c}$ should be smaller than the moment resistance of the reinforced concrete cross-section $M_{Rd,c}$ (i.e. $M_{Ed,kc} \leq M_{Rd,c}$ and $M_{Ed,c} \leq M_{Rd,c}$).

As mentioned at the beginning of this section, the moment capacity of each cross-section should be calculated by accounting for linear cross-sectional compatibility and considering that the flexural reinforcement A_{sl} reaches its yield strength ($\sigma_{sl} = f_{ysd}$). The forces in the shear key can be computed considering equivalent reinforcement A_{aft} for the top flange, A_{aw} for the web and A_{afb} for the bottom flange (Figure 6.7b).

The moment resistance can be calculated with the expressions given below for a composite cross-section $M_{Rd,k}$ and for a non-composite cross-section $M_{Rd,c}$, where the depth of the corresponding compression zones c_k and c_c should be determined separately for each case, respectively. The notation <x> denotes a Macaulay bracket, indicating that if the value within the brackets is negative, it is considered zero, whilst if the term is positive, the assessed value is considered. The parameter λ defines characteristics of stress distribution in the compression zone. This can be taken as $\lambda = 0.8$ for normal concrete ($f_{ck} \leq 50$ MPa). Other methods to assess the cross-sectional capacity can be used, as long as they follow the assumptions and principles described above. The assessment of the depth of the composite compression zone is carried out by obeying linear strain compatibility, equilibrium, and constitutive conditions within the cross-section. This can be achieved by using an iterative procedure in which the initial depth of the compression zone $c_{k,0} = \lambda_c c_c$ and $\lambda_c \geq 1.0$, so that eventually equilibrium is satisfied and $c_k = c_{k,0}$.

$$M_{Rd,k} = f_{ysd}\left\{ A_{sl}\left(d - \frac{c_k}{2}\right) + \Sigma\left[A_{aij}\left\langle (d_{aij} - c_k)(d_{aij} - c_k/2)\right\rangle\right] / (d - c_k)\right\}$$

(6.1)

$$M_{Rd,c} = f_{ysd}A_{sl}\left(d - \frac{c_c}{2}\right)$$

(6.2)

$$c_k = \frac{f_{ysd}\left(A_{sl} + A_{aft}\left\langle \dfrac{d_{aft} - c_{k,0}}{d - c_{k,0}}\right\rangle + A_{aw}\left\langle \dfrac{d_{aw} - c_{k,0}}{d - c_{k,0}}\right\rangle + A_{afb}\left\langle \dfrac{d_{afb} - c_{k,0}}{d - c_{k,0}}\right\rangle\right)}{\lambda f_{cd}b_c}$$

(6.3)

$$c_c = \frac{f_{ysd}A_{sl}}{\lambda f_{cd}b_c}$$

(6.4)

where $\langle x \rangle : (< 0 = 0; \geq 0 = x)$

The tension longitudinal reinforcement in the composite region of the hybrid connection should be greater than 0.4% and less than 2.0%, considering the bare concrete section without the contribution of the shear key. The longitudinal reinforcement should be continuous above the column support. This can be achieved by drilling through the column flanges, or by passing the rebars around the column allowing for adequate spacing such that the concrete can be placed and compacted satisfactorily for the development of adequate bond. To avoid congestion of reinforcement and localized stress concentrations, lapping of bars, welding and mechanical devices should not be used within the composite region and at the interface. Laps, welding, and mechanical couplers are allowed only in the non-composite region accordance with Eurocode 2 for reinforced concrete components.

6.3.2 Shear strength

The following section introduces a method to assess the shear strength of composite and reinforced concrete sections in the two regions of the hybrid connection: the composite and the transition regions, respectively. The shear capacity should be calculated in all distinct regions of the hybrid connection described above, and the design of the non-composite regions should follow typical procedures for reinforced concrete components outlined in Eurocode 2. The shear design of the composite and transition regions is based on the Eurocode principles with due account for the geometrical characteristics of the hybrid connections. Note that the rules described below were validated for beam width-to-depth ratios in the range of 0.75–1.25. The design of sections outside these ranges should be accompanied by detailed investigations. The composite region of the hybrid connections should be checked for vertical and longitudinal shear resistance, and the effects of local buckling may be neglected for a steel section fully encased in accordance with Eurocode 4 specifications. In all sections, the shear resistances (V_{Rd}) should be greater than the design shear actions (V_{Ed}).

In the composite region, at least a minimum cover of reinforced concrete shall be provided to ensure the safe transmission of bond forces, the protection of the steel against corrosion, and spalling of concrete. The concrete cover should not be below 40 *mm* and in accordance with the Eurocode 4 specifications, but also obey the procedures described in Eurocode 2 for longitudinal reinforcement. The tensile longitudinal reinforcement that may be used in the design should be greater than 0.4% but less than 2.0%, considering the bare reinforced concrete section.

To achieve some degree of composite action, the composite region should be provided with transverse reinforcement in the form of adequately anchored closed stirrups (shear links), perpendicular to the centreline along the length of the shear key. The transverse spacing of the legs between two consecutive stirrups should be less than the minimum of 75% of the effect shear depth in the interface region of the hybrid connection, the shear key depth, and the stirrup spacing in the interface and non-composite region, i.e. $s_{t,max} \leq min$ ($0.75 \times d_0$, d_a, $s_{t,nc}$). The distance s_0 between the column face and the first stirrups should be between 30% and 70% of the effective shear depth ($s_0 = 0.3\text{--}0.7 \times d_0$). Although reinforced concrete sections without transverse reinforcement, outside of the shear key length, may be considered, the amount of transverse reinforcement in the transition region should not be less than that in the composite region of the hybrid connection.

It is worth pointing out that the validation of the shear design procedures for the transition zone considered a cumulative contribution of the concrete and transverse reinforcement to the shear resistance yet, for harmonization with the current and revised versions of Eurocode 2, the shear design considers two distinct cases. The first case is for components not requiring transverse reinforcement (i.e. the concrete shear capacity V_{Rdc} is greater than the design shear force V_{Ed}), and the second is when the concrete shear capacity is smaller than the shear demand ($V_{Rdc} < V_{Ed}$). In the latter, the shear resistance of a reinforced concrete section in the transition zone is based on the contribution of the transverse reinforcement only, and it is limited by the strut-crushing capacity.

6.3.2.1 Composite region

6.3.2.1.1 Vertical shear

In one-way hybrid connections to reinforced concrete beams, with beam width-to-depth ratios within the ranges described above, the contribution of concrete to the vertical shear capacity can be ignored as recommended by Eurocode 4. Partial contribution of transverse reinforcement in the form of closed stirrups can be considered, only if the transverse reinforcement area is greater or equal to that in the transition region. In this situation, the resistance to vertical shear $V_{pl,a,Rd}$ should be taken as the resistance of the structural steel section, as per Eurocode 4 recommendations $V_{pl,a,Rd}$, and of the transverse reinforcement $V_{Rd,s}$, as per Eurocode 2 guidelines.

$$V_{Rd,k} = V_{pl,a,Rd} + V_{Rd,s,k} \qquad (6.5)$$

$$V_{pl,a,Rd} = \frac{A_{a,v}\left(f_{yak}/\sqrt{3}\right)}{\gamma_{M0}} \qquad (6.6)$$

$$A_{a,v} = \max\left(A_a - 2b_a t_f + \left(t_w + 2r\right)t_f, \eta h_w t_w\right) \quad \text{For rolled I and H sections}$$

loaded parallel to the web, the shear area Aa,v may be taken as: (6.7)

Where the design vertical shear V_{Ed} exceeds half the design plastic resistance $V_{pl,a,Rd}$ of the structural steel section to vertical shear, allowance should be made for its effect on the moment resistance. The influence of the vertical shear on the resistance of bending may be calculated by using a reduced yield strength by multiplying the shear key design yield strength f_{yad} by $(1-\rho)$.

$$(1 - \rho)f_{yad} \tag{6.8}$$

$$\rho = \left(\frac{2V_{Ed}}{V_{pl,a,Rd}} - 1 \right)^2 \tag{6.9}$$

The contribution of the transverse reinforcement to the shear capacity $V_{Rd,s,k}$ of the composite cross-section is given by Equation (6.10), and it is limited by crushing in the concrete panel zone. The shear resistance $V_{Rd,s,k}$ is a function of the shear reinforcement area A_{sw}, the lever arm in the composite cross-section $z_k=d-c_k/2$, or $z_k=0.75d$ in a simplified manner, the yield strength of the shear reinforcement f_{ywd}, the stirrup spacing s_w, and the inclination of the critical strut assumed $1\le \cot\theta \le 1.5$ for hybrid connections, based on experimental evidence (Bompa and Elghazouli, 2015; Moharram et al., 2017a).

The crushing capacity, represented by the right-hand quantity of the $V_{Rd,s,k}$ equation, depends on the compressive strength reduction factor which can be assumed as $\nu=0.5$, concrete compressive design strength f_{cd}, thickness of the concrete panels (b_c-b_a), as shown in Figure 6.8, and the lever arm in the composite region z_k.

$$V_{Rd,s,k} = \frac{A_{sw} \cdot z_k \cdot f_{ywd} \cdot \cot\theta}{s_w} \le 0.5\nu f_{cd}(b_c - b_a)z_k \tag{6.10}$$

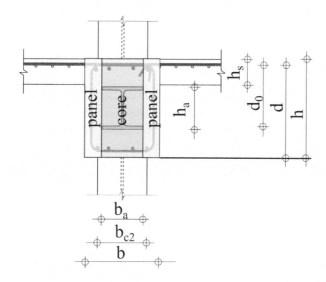

Figure 6.8 Details of the main shear transfer cross-sectional regions

6.3.2.1.2 Longitudinal shear

The longitudinal shear $V_{Rd,L,k}$ at the interface between concrete and steel should be verified where it is caused by transverse loads and/or end moments according to Eurocode 4 procedures. The design shear strength of $\tau_{Rd}=0.30$ N/mm^2 for unpainted and free from oil, grease and loose scale or rust can be assumed. Allowance for higher values of τ_{Rd} can be made based on experimental evidence or using the values reported in Sections 2.4 and 2.5.. For concrete covers greater than 40 mm, the design shear strength τ_{Rd} can be multiplied by β_c. The parameter A_{vL} is the contact area of the steel profile, as obtained from manufacturer tables or conservatively assumed as $A_{vL}=4b+2h-2t_w-4r-4t_f$, whilst l_v is the embedment length of the shear key.

$$V_{Rd,L,k} = A_{vL}\beta_c\tau_{L,Rd}l_v \tag{6.11}$$

$$\beta_c = 1 + 0.02c_z\left(1 - \frac{40}{c_z}\right) \leq 2.5 \tag{6.12}$$

To achieve a full-strength connection between the shear key and concrete, shear studs can be welded to the compression flange of the steel insert. This can enable full interaction between the two components. This hypothesis requires a definition of the design longitudinal shear $V_{Ed,L,k}$ at the interface between the steel flange and the concrete. This is undertaken by defining an effective interacting concrete section and a design compression stress in that concrete. It is considered that: (i) the effective concrete interacting with the shear key is equal to the width of the insert b_a, (ii) the total effective width of the steel insert is $l_v/8 + b_a$, (iii) the concrete thickness should be calculated using either the principles described above for flexural assessments or by assuming a plastic distribution in the cross-section. The design should be carried out using Eurocode 4 principles.

The number of welded studs required to ensure a full-strength connection n_{ws} is the ratio between the design longitudinal shear $V_{Ed,L,k}$ and the design resistance of a welded shear stud P_{Rd}. The latter is the minimum of two conditions: shear stud yielding or concrete crushing. The shear stud yielding is a function of the ultimate material strength in the shear connector f_u and its diameter d_{ba}, whilst the crushing strength is a function of α (a parameter depending on the stud aspect ratio), diameter d_{ba}, and concrete material properties (f_{ck} and E_c). Both conditions are divided by the partial safety factor $\gamma_a=1.25$.

$$V_{Ed,L,k} = \lambda f_{cd}c_{k,pl}\left(l_v/8 + b_a\right) \tag{6.13}$$

$$n_{ws} \geq V_{Ed,L,k}/P_{Rd} \tag{6.14}$$

$$P_{Rd} = \min\left(0.8f_u\pi d_{ba}^2/4; 0.29\alpha d_{ba}^2\sqrt{f_{ck}E_c}\right)/\gamma_a \tag{6.15}$$

6.3.2.2 Transition regions

6.3.2.2.1 Transition regions not requiring design shear reinforcement

The design shear equations for transition regions in which the design shear force is smaller than the concrete contribution to shear strength ($V_{Ed,kc} < V_{Rd.kc}$) are largely based on the fundamentals of the revised Eurocode 2 (CEN, 2018). If non-shear-reinforced configurations are chosen through design, the transverse bar detailing for the composite region must be extended to the composite-to-reinforced concrete interface within the transition region. The contribution of the shear key ρ_a to the combined reinforcement ratio ρ_{tot} is multiplied with a factor that depends on the embedded length l_v and component effective shear depth d_0, denoted as the distance between the top side of the bottom flange of the shear key and the centroid of the longitudinal reinforcement (see Figure 6.7).

The combined reinforcement ratio ρ_{tot} considers the contribution of the longitudinal reinforcement ratio ρ_l and the shear key reinforcement ratio ρ_a. The parameter A_{sl} is the total area of longitudinal reinforcement that complies with bond requirements outlined in Eurocode 2, and A_a is the shear key cross-section area assessed in accordance with the requirements described in Section 6.34.

In typical design situations, non-composite components connected to steel columns by means of shear keys are expected to have very low levels of axial load. Hence, the influence of a compression load on the shear capacity is not accounted for in the design.

The shear resistance of components not requiring shear reinforcement $V_{Rd,c,kc}$ is given below. This is a function of the combined reinforcement ratio ρ_{tot}, characteristic concrete compressive strength f_{ck} in MPa, the reference value of the roughness of the critical shear crack d_{dg} (where d_{g0} is the reference aggregate size equal to 16 mm for normal concrete, and d_g maximum aggregate size in the concrete), d_0 is the effective shear depth, and b_c is the width of the component.

$$V_{Rd,c,kc} = \frac{0.6}{\gamma_c}\left(100\rho_{tot}f_{ck}\frac{d_{dg}}{d_0}\right)^{1/3} b_c d_0 \tag{6.16}$$

$$\rho_{tot} = \rho_l + \left(0.2 \cdot l_v/d_0\right)^3 \rho_a \tag{6.17}$$

where $\rho_a = A_a / \left(b_c d_a\right)$

and $\rho_l = A_{sl} / \left(b_c d\right) \leq 0.02$

$$d_{dg} = d_{g0} + d_g \tag{6.18}$$

$$d_0 = d - d_a + h_a/2 - t_f \tag{6.19}$$

6.3.2.2.2 Transition regions requiring design shear reinforcement

When the design shear force is greater than the concrete contribution to shear strength ($V_{Ed,kc} \geq V_{Rd,kc}$), the shear resistance is given by the shear reinforcement at yielding $V_{Rd,s,kc}$, and is limited by concrete crushing. The latter considers that the inclination of the critical strut is fixed at 45° in agreement with Eurocode 2 provisions.

Based on limit analysis, the position of the neutral axis in test components (Bompa and Elghazouli, 2015) was found to be in the region of the bottom flange of the composite region and in a lower position in the conventional RC case. As observed in the tests, failure occurred at the interface region where the neutral axis is found to be mid-way between the two. The lever arm accounted for in shear calculations in typical reinforced concrete components (i.e. $z=0.9d$) is slightly higher than that in the transition region of hybrid connections described herein.

A representation of the lever arm corresponding to 75% of the effective bending depth in the composite region and 80% in the transition region, leads to a more satisfactory estimation of the shear resistance of hybrid connections within the interface region requiring transverse reinforcement. The shear resistance of transition regions provided with transverse reinforcement $V_{Rd,s,kc}$ uses the same parameters described above, noting that the lever arm in the transition region, z_{kc} is $z_{kc}=d-c_{kc}/2$ or in a simplified manner $z_{kc}=0.80d$, where $c_{kc}=(c_c+c_k)/2$.

In the transition region, the governing strut is supported on the bottom flange of the shear key. Therefore, the vertical projection of the strut reduces when compared to a typical reinforced concrete component. Similarly to the case when the contribution of the transverse reinforcement is estimated, the lever arm z_{kc}, as typically used for reinforced concrete components, is reduced. In the case of shear failure resulting from strut crushing, the "hybrid" shear lever arm d_0 becomes the distance between the top face of the bottom flange of the shear key and the centroid of the longitudinal reinforcement (Figure 6.9). Hence, the crushing capacity is limited by the strut support offered by the shear key flanges (b_a-t_w).

$$V_{Rd,s,kc} = \frac{A_{sw} \cdot z_{kc} \cdot f_{ywd} \cdot \cot\theta}{s_w} \leq 0.5 v f_c (b_a - t_w) d_0 \qquad (6.20)$$

In both the composite and transition regions, the critical shear crack that can be chosen between $1 \leq \cot\theta \leq 1.5$ for hybrid connections, based on experimental and numerical evidence (Bompa and Elghazouli, 2015; Moharram et al., 2017a). The governing shear crack inclinations observed in experiments varied between 36 to 44 degrees. In design, crack inclinations should be determined starting from elastic stress distribution (45°), which will give the most conservative result since the number of transverse bars intersected by the crack is minimal. Alternatively, this can be addressed by accounting

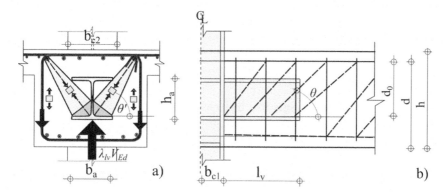

Figure 6.9 Strut development schemes: a) cross-section, b) within connection length

for a compatible strut-and-tie mechanism, considering that the governing strut is supported by the tip of the shear key and joins the first stirrup located at a distance equal or higher than $0.75d$ from the tip of the shear key.

For concentrated loads applied near the shear key tip, direct struts develop from the load application point down to the flanges of the shear key. Following an extensive set of parametric studies adopting a wide range of material and geometrical properties (Moharram, 2018), it was observed that a direct strut mechanism develops, in most cases, at shear span-to-depth ratios around $a_v/d < 0.66$. The shear span is defined as the distance between the load application point and the tip of the shear key. In contrast with reinforced concrete beam design, when allowance is made for strength enhancement due to direct strut support for $a_v/d \leq 2.5$, the behaviour of hybrid connections is fundamentally different. In such connections, there are three struts transferring loads from the non-composite region to the column by means of a shear key, whilst in conventional reinforced concrete, the force transfer is made through a single-governing strut having the width of the beam.

In hybrid connections, the three struts underlying the direct strut mechanism contribute to a different extent to the force transfer. A main critical strut bearing against the bottom flange of the shear key has the highest contribution. The force transferred by this strut is limited by the width of the shear key. A second strut bears against the top flange of the shear key, whilst a third strut, complements the system equilibrium. The latter is connected to the beam compression zone, having a minimum contribution to the ultimate capacity of the direct mechanism (Figure 6.3). To account for the loads taken to the support through the direct strut for $a_v/d_0 < 0.70$, the shear resistance can be enhanced by a factor $\lambda_\beta = 2.5 \times d_0/a_v$ that is based on typical assumptions in Eurocode 2, which consider an inverse reduction β factor applied to the design shear force.

The *non-composite region* of the hybrid connection should be designed according to specific guidance for reinforced concrete available in Eurocode 2, considering the constraints imposed by the design of the composite and transition regions described above.

6.3.3 Shear key design for hybrid connections to one-way components

The design of the composite sections of hybrid connections, incorporating the shear keys should follow the guidance provided in the sections above for bending and shear. Using the principles described at the beginning of the section, the design of the shear key cross-section can be assessed by means of linear strain compatibility in the composite cross-section (Figure 6.7). This can be achieved by simply assuming the entire steel insert is represented by a single reinforcement bar located in the geometrical centre of the shear key.

The stress in a reinforcement component and its contribution to the moment resistance can be determined using fundamental constitutive relationships, and by assuming a cracked second moment of inertia of the composite section. The latter can be determined using established procedures by considering the cumulative contribution of the reinforcement elements as well as the cracked moment of inertia of the reinforced concrete section. The cracked moment of inertia of the concrete section depends on the position of the neutral axis and can vary between 30–70% of the un-cracked cross-section. Based on experimental results, at ultimate, an average ratio of $c_k/d{=}0.4$ was obtained.

The cross-section of the shear key should be determined from the highest design moment $M_{Ed,k}$ and shear force $V_{Ed,k}$ within the composite region. The second moment of inertia I_a of the shear key can be obtained by accounting for the linear strain compatibility and yielding of the foremost reinforcement material, whilst the cross-sectional shear area $A_{a,v}$ from the assumption that all design shear $V_{Ed,k}$ is carried out by the shear key. Both conditions need to be satisfied. The definition of the effective shear area of a steel profile should follow the definitions available in Eurocode 3. A reduced shear design force $V_{Ed,k}$ can be considered for determining $A_{a,v}$ by undertaking detailed investigations, assuming that part $V_{Ed,k}$ is transferred by the transverse reinforcement in the concrete panels.

$$I_a \geq \frac{M_{Ed,k}\left(d - c_k\right)}{10 f_{ysd}}$$

(6.21)

$$A_{a,v} \geq \frac{V_{Ed,k}}{f_{yak}/\sqrt{3}}$$

(6.22)

Note that when the design shear $V_{Ed,k}$ exceeds half the design plastic resistance $V_{pl,a,Rd}$ of the structural steel section to vertical shear, allowance should be made for its effect on the moment resistance. The influence of the vertical shear on the moment resistance may be calculated by using a reduced yield strength using principles of Eurocode 3 and 4, as described in Section 6.3.2.

The length of the steel insert l_v should be determined as the maximum value of the four conditions described below. The first condition assumes that the load transfer from the component to the column is made through the compression field that is supported on the bottom flange of the shear key. Hence, the embedment length $l_{v,1}$ results from the limit condition assuming that the width of the shear key bottom flanges $(b_a$-$t_w)$ can transfer the total force through a strut of a strength equal to $0.5 \nu f_{cd}$.

Moreover, the load transfer through struts is limited by the capacity of transverse reinforcement connected to the compression field. The second condition to determine the embedment length $l_{v,2}$ is given below. Noting that the support length of the strut should be greater than the horizontal projection of the strut to allow force transfer to a sufficient amount of transverse bars, the embedment length should $l_{v,3}$ be greater than the horizontal strut projection. Additionally, the shear key length should be greater than the shear key depth h_a.

$$l_{v,1} \geq \frac{V_{Ed,k}}{0.5\nu\left(b_a - t_w\right)f_{cd}} \tag{6.23}$$

$$l_{v,2} \geq \frac{V_{Ed,k}s_w}{A_{sw}f_{yw}} \tag{6.24}$$

$$l_{v,3} \geq d_0 \cot\theta \tag{6.25}$$

$$l_v = \max(l_{v,1}, l_{v,2}, l_{v,3}, h_a) \tag{6.26}$$

Using the sizing principles described above, the shear design of the shear key is very likely to be conservative with the shear capacity of the steel insert $V_{pl,a,Rd}$ superior to the shear design force $V_{Ed,k}$. As the pre-sizing in bending was undertaken considering the shear key response in a composite cross-section, its cross-section must be verified to resist the bending produced by the reaction forces.

It has been observed experimentally that the shape of the distribution of the reaction force changes with the length of the shear key (Bompa and Elghazouli, 2015; Moharram et al., 2017b). Regardless of the shear key length, at least one-third of the reaction force is transferred within $h_a/2$ from the column face, whereas the remaining force is transferred within

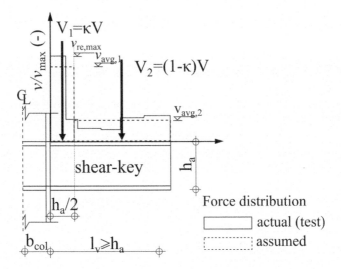

Figure 6.10 Reaction force distribution for shear keys in hybrid connections to beams

l_v-$h_t/2$. It results in a double-rectangular distribution with peak force at the face of the column $v_{Ed1} \geq v_{Ed2}$. The required moment resistance $M_{a,Ed}$ from the force reactions must be smaller than the elastic moment capacity of the steel insert $M_{a,el,Rd}$. The latter can be assessed using available procedures in Eurocode 3. The design bending moment, assessed from force reaction, may be calculated by accounting for the double-rectangular reaction force distribution, as illustrated in Figure 6.10.

$$M_{a,Ed} = \frac{V_{Ed}}{3}\left(l_v + \frac{h_a^2}{8l_v}\right) \tag{6.27}$$

6.3.4 Validation and practical recommendations

The formulations described above have been validated on a series of 14 test specimens and around 200 non-linear numerical models (Bompa and Elghazouli, 2015; Moharram et al., 2017a, 2022). All geometries replicated the joint region of a hybrid moment frame consisting of steel columns and reinforced concrete beams. The main parameters varied within the experiments were the embedded length of the shear key (embedded length-to-steel component depth l_v/h_a=1.0–3.6), the presence of transverse reinforcement (13 specimens had stirrups with ρ_w=0.19% and one was without), the flexural reinforcement ratio (ρ_l=0.3% - 1.21%), and the stiffness ratio between the component and the shear key. This ratio is represented by $\eta=E_c I_c/E_a I_a$ which is dependent on the elastic concrete modulus E_c, the elastic moment of inertia of the concrete cross-section I_c, the elastic steel modulus E_a obtained from material tests and the moment of inertia of the shear key I_a. After undertaking validation of the modelling procedures on the experiments mentioned above, as well as an established test series on reinforced concrete beams (Vecchio and Shim, 2004; Moharram et al., 2017a), a set of numerical models were constructed. To complement the experimental ranges, the main parameters varied within the parametric investigations were concrete compressive strength (f_c=16–50 MPa), reinforcement ratio (ρ_l=0.3–1.9%), shear key section type HEB180-HEB220, embedment length-to-depth ratio l_v/h_a=0.5–10.0, and various transverse reinforcement ratios by varying the stirrup diameter (8–12 mm) and stirrup spacing (75–200 mm). Note that the parametric studies were primarily used to validate the shear strength design models from this section.

From the total number of experiments on hybrid connections, four elements have been provided with low reinforcement ratios failing in flexure, as schematically presented in Figure 6.11a. The remaining components had relatively high reinforcement ratios and had shear-controlled failures (Figure 6.11b). As noted above, a hybrid connection of steel columns to RC beams has three regions (composite, transition, and non-composite), and two types of cross-sections (composite and reinforced concrete). Using the equations above, considering the component's geometry and test material properties, as well as the specific test loading scheme, the flexural strength was assessed at the column face (composite region – composite cross-section) and the interface (transition region – reinforced concrete cross-section). Note that in all assessments, the test material properties have been used and material safety factors equal to unity (γ_i=1.0).

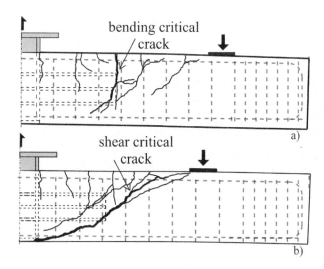

Figure 6.11 Crack details of test specimens failing in a) bending, b) shear

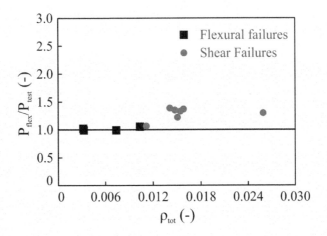

Figure 6.12 Validation of the flexural strength design approach for hybrid connections to beams

Figure 6.12 shows the relationship between the test-to-flexural strength ratio (P_{test}/P_{flex}) and the combined reinforcement ratio (ρ_{tot}). Values of $P_{flex}/P_{test}{\leq}1.00$ indicate flexural failures, whilst $P_{flex}/P_{test}{>}1.00$ represents shear failures. As indicated in the figure, specimens with low reinforcement ratios (marked with black squares), in which the strain gauges located on the tension longitudinal reinforcement showed values above the yield limit, are situated around the 1:1 line. Specimens that did not show flexural yielding and failed in shear, marked with grey circles, are above the same line. This shows that the proposed approach for assessing the bending resistance at various cross-sections of the hybrid connection is adequate.

Extensive validation of the shear strength expressions was undertaken based on the test results and the non-linear simulations as described above. Note that for the comparative assessments the material test strengths, as well as material safety factors equal to unity ($\gamma_i{=}1.0$), were used. The experimental results provided insight into the influence of various shear transfer mechanisms including transverse reinforcement, compressive zones, residual tensile stresses, aggregate interlock, and dowel action, in addition to the interfacial bond between the steel profile and concrete, to the shear strength of such systems. Based on these assessments, modifications to the current Eurocode 2 model were made to incorporate the shear key influence on the shear response (Bompa and Elghazouli, 2015). To ensure consistency in formulation with the punching shear design model described below, as well as for harmonizing the proposed design with the revised European concrete design provisions, the model is presented here in an updated from.

The failure kinematics of all tests with relatively high reinforcement ratios $\rho_l{\geq}1.0\%$ and numerical models $\rho_l{\approx}1.1\%$, was characterized by the development of critical shear crack in the transition region. The crack intersected

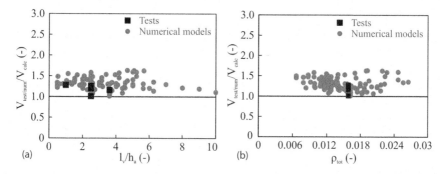

Figure 6.13 Validation of shear design model for hybrid connections to beams: a) embedment length, b) combined reinforcement ratio

the shear key around its tip and passed through several stirrups, depending on the crack inclination. In these assessments, the critical crack inclination was using the ranges imposed by the design model ($cot\theta=1.5$) and within the values observed in the experiments. Figure 6.13 shows the relationships between the predicted-to-test or numerical shear strength versus shear key length-to-depth ratio l_t/h_a (Figure 6.13a) and the combined reinforcement ratio ρ_{tot} (Figure 6.13b). Values of $V_{test/num}/V_{calc}\geq1.0$ indicate safe estimates, whilst $V_{test/num}/V_{calc}<1.0$ are unconservative.

It is worth pointing out that the validation was undertaken by assuming a cumulative contribution of the concrete and transverse reinforcement to the total shear resistance, as observed from detailed measurements of specimen kinematics (Bompa and Elghazouli, 2015). However, for practical design, the application of the shear design model should follow the guidance described above that is conforming with the Eurocode 2 approach.

The predicted results show good agreement with those obtained from tests and simulations for a wide range of geometries with various shear key embedment lengths (l_t/h_a=1.0–10.0), with stiffness ratios that involve relatively rigid behaviour of the shear key, using normal concrete, and provided with intermediate RC flexural reinforcement ratios. The average $V_{test/num}/V_{calc}$=1.30, a coefficient of variation of 0.10, and a 5% percentile of 1.10, indicate safe estimates across all geometry and material ranges. More conservative estimations could be obtained if design safety factors are used.

The rationale behind the shear key force distributions suggested in Section 6.3.4 results from an in-depth analysis of the test results. For all the material and geometrical characteristics examined in the experiments, the embedded shear key was relatively strong, with no local failures in the steel insert taking place at the ultimate load. To better understand the shear force distribution and reactions at the shear key, parametric investigations showed that for relatively flexural reinforcement ratios (e.g. ρ_l=0.3%), the shear key takes a higher amount of force than in the case of relatively high reinforcement ratios (ρ_l=1.1–1.9%) (Moharram, 2018). In the former case, about 80% of the total force is transferred by the shear key, whilst it varies in the second but can be as low as 40%. As expected, this highly depends on the shear key stiffness and overall cross-section rigidity. In design, it can be conservatively assumed that all the force is transferred by the shear key, assuming that both in bending and shear, the shear key is carrying all the load, in agreement with Eurocode 4 guidelines.

Besides the interaction between the shear key cross-section and flexural reinforcement area, the behaviour of the hybrid connection is highly dependent on the shear key embedment length, which dictates the effective reaction point of the loads transferred from the moment span. A detailed analysis of the shear distribution recorded from the 14 specimens from the experimental programme showed that the shear force distribution and its centre of application depended on the shear key stiffness and length. More rigid and short shear keys had the effective point of application moving towards the cantilever end, whilst more flexible configurations with the reaction force towards the column face. In a simplified approach, for short shear keys, the doubly rectangular distribution, described above, can be replaced by a force acting at the shear key tip, assuming a rigid shear key.

Close inspection of the preliminary sizing equations for the shear key embedment length, showed that assuming that the insert length is assessed solely from the crushing capacity of the struts, leads to extremely short lengths (below $h_a/2$). This might be unrealistic since the increase in concrete strength leads to shorter embedment lengths. Higher concrete strengths will allow a higher amount of load transferred through the strut which leads to higher deformability of the shear key. These two seem to be in contradiction. Alternatively, the assumption that the pressure resulting from the reaction force equals the strut stress at the yield of stirrups offers more realistic insert lengths, similar to those used in experiments.

For hybrid connections to one-way components, relatively short embedment lengths of the shear key seem to be more effective from the practical point of view. A ratio of l_s/h_a=1.0 appears to be sufficient to ensure a smooth transfer of forces between the reinforced concrete beam and steel column, and a stiff response of the shear key. The structural response of a hybrid component with a short shear key tends to resemble that of a typical reinforced concrete component, hence may be a desirable solution in practice. Longer shear keys provide supplementary strut support and an extension of the composite region, translating the weak regions further away from the column face. Depending on the moment and shear demand in the vicinity of the connection, higher shear key length-to-depth ratios l_s/h_a>1.0 can be considered. Relatively longer shear keys with l_s/h_a>3.0 are likely to act compositely with the surrounding concrete, with the overall connection behaviour moving away from that of a typical RC component.

In hybrid connections between steel columns and primary seismic beams, the shear design forces shall be determined in accordance with the capacity design rules of Eurocode 8 for reinforced concrete beams (CEN, 2004c). It is expected that in beams incorporating shear keys, the plastic hinge will develop within the transition region for positive and negative directions of seismic loading (Figure 6.2). Hence the maximum and minimum shear force should be calculated at the composite-to-reinforced concrete interface, corresponding to maximum positive and maximum negative interface moments $M_{i,d}$.

The moments at the composite-to-reinforced concrete interface may be determined as the product of an overstrength factor (γ_{Rd}=1.0 for ductility class medium DCM), the design moment resistance of the reinforced cross-section at the interface $M_{Rd,c}$ and the ratio of the sum of moments framing the connection ($\sum M_{Rc}/\sum M_{Rb} \geq 1.0$, where $\sum M_{Rc}$ is the sum of steel column section moment resistances, and $\sum M_{Rb}$ is the sum of the reinforced concrete section moment resistances). The column moment resistance should correspond to the accidental design situation. The flexural and shear resistances should be assessed using the equations in Section 6.3.1 and 6.3.2, respectively. The detailing of the transition region, in which the plastic hinge is expected to develop, should follow the guidance in Eurocode 8 for primary seismic reinforced concrete beams. To satisfy the local ductility requirements in critical regions of primary seismic beams, the detailing requirements and material qualities stipulated in Eurocode 8 should be adopted. The transverse reinforcement detailing for the transition region, corresponding to the critical region for ductile design, should extend to the column face.

6.3.5 Design example

6.3.5.1 Description of the structure

The design of a one-way moment frame floor system is required in which a hybrid structural system made of steel columns and reinforced concrete beams is chosen. The force transfer is made through a short steel insert, referred to as the shear key, welded to the steel column and fully integrated into the reinforced concrete beam.

Exposure class XC1 (Concrete inside buildings with low air humidity)
Structural class S4, for C50 -> reduce with class 1, results in S3

$$L_x = 16.5\,m \quad L_y = 9.50\,m$$

$$b_{c1} = 800\,mm \quad b_{c2} = 600\,mm$$

The design bending moments and shear forces are obtained through an elastic FE analysis using rigid connections. The design bending moment M_{Ed}=4569 kNm and the design shear force V_{Ed}=1645 kN.

Assuming a span-to-depth ratio L/d=26 for continuous components, an initial effective bending depth of d_{init}=630 mm can be considered. Moreover, as an initial estimate of the shear key length, consider $d_{init}=l_{v,init}=630\ mm$. Accounting for the column size as noted above,

the bending moment and the shear force at the column face in the composite region are $M_{Ed,k}=3307\ kNm$ and $V_{Ed,k}=1565\ kN$,

the bending moment and the shear force at the composite-to-RC interface in the transition region are $M_{Ed,kc}=1931\ kNm$ and $V_{Ed,kc}=1438\ kN$,

Material properties:

Concrete grade C50 $f_{ck} = 50\ MPa\ \gamma_c = 1.5\ f_{cd} = 33.3\ MPa\ d_g = 20\ mm$

Longitudinal steel grade B500S $f_{ysk} = 500\ MPa\quad f_{ysd} = 435\ MPa$

Transverse steel grade B500S $f_{ywk} = 500\ MPa\quad f_{ywd} = 435\ MPa$

Structural steel Grade S460 $f_{yak} = 460\ MPa\quad \gamma_{M0} = 1.00\quad f_{yad} = 460\ MPa$

Preliminary sizing:

The slab depth h_s is given by the short span minus the width of the beam, assumed in this stage as: $h_s = \left(L_y - b_{c,init}\right)/35 = (9.5-1.0)m/35 \cong 240\ mm$

From the initial design using $L/d=26$, the initial beam depth considering the concrete cover is

$$h_{init} = d_{init} + 70\ mm = 700\ mm$$

Beam reinforcement design:

An iterative procedure leads to a total tension reinforcement amount consisting of $6\phi40\ mm$ with an amount of $A_{sl} = 7540\ mm^2$.

From technological requirements, the width b of the beam is limited by the amount of reinforcement and minimum spacing between bars $s_{l,min}$

$$b_c = \max(b_{c,init,a}, b_{c,init,b}) \cong 700\ mm$$

$$b_{c,init,a} = p_0 + n_{bl} \times d_{bl} + \left(n_{bl} - 2\right) \times s_{l,min} + 2 \times d_{bw} + 2 \times c_{nom}$$

$$b_{c,init,b} = 140 + 6 \times 40 + 4 \times 40 + 2 \times 12 + 2 \times 40 = 644\ mm$$

$$b_{c,init,a} = b_{c2} + 100\ mm = 700\ mm$$

p_0 is the average allowed drilling distance in the column flanges

$$s_{l,min} = \max\left(d_{bl,max} = 40\ mm,\ d_g + 5\ mm = 25\ mm\right) = 40\ mm$$

As indicated, both the beam depth and width $h=b_c=700\ mm$ $(h/b_c=1.0)$, hence not obeying typical beam h/b_c ratios. To benefit from a greater lever arm in bending, the beam depth is increased to $h=100\ mm$. Following an iterative procedure, the cross-section of the reinforced concrete beam is chosen as:

$$b_c \times h = 700\ mm \times 1000\ mm$$

The plastic moment of the concrete cross-section is given by:

$$M_{Rd,c} = f_{ysd}A_{sl}\left(d - \frac{c_c}{2}\right) = 435 \times 7539\left(928 - \frac{165}{2}\right) = 2754\,kNm$$

in which the depth of the compressive zone is:

$$c_c = \frac{f_{ysd}A_{sl}}{\lambda f_{cd}b} = \frac{435\,MPa \times 7539\,mm^2}{0.8 \times 33.3\,MPa \times 700\,mm} = 175.6\,mm$$

And the bending effective bending depth is:

$$d = h - c_{nom} - d_{bw} - d_{bl,max}\,/\,2 = 928\,mm$$

$$c_{min,sl} = \max\left(c_{min,b} = 40\,mm,\; c_{min,dur} = 10\,mm,\; 10\,mm\right) = 40\,mm$$

$$c_{min,sw} = \max\left(c_{min,b} = 12\,mm,\; c_{min,dur} = 10\,mm,\; 10\,mm\right) = 12\,mm$$

$$c_{nom} = \max(c_{min,sl} - c_{min,sw},\, 40\,mm) \cong 40\,mm$$

The moment capacity-to-demand, considering the bending resistance of the reinforced concrete section, is $M_{Rd,c}/M_{Ed,kc}$=1.43 in the transition region at the composite-to-RC interface and $M_{Rd,c}/M_{Ed,k}$=0.83 at the column face. This is satisfactory, as the shear key contribution to the bending capacity has not been considered yet. The flexural reinforcement ratio ρ_l=1.16% is within the limits imposed by the codified procedures.

$$M_{Rd,c} = 2754\,kNm > M_{Ed,kc} = 1931\,kNm$$

$$M_{Rd,c} = 2754\,kNm < M_{Rd,k} = 3307\,kNm$$

Shear key design
  Cross-section design for shear
  The required shear plastic resistance of the shear key is:

$$A_{a,v} \geq \frac{V_{Ed,k}}{f_{yak}\,/\,\sqrt{3}} = 6193\,mm^2$$

This results in a European standard HEB500 profile satisfying the demand:

$$A_{a,v} = 8982\,mm^2$$

Cross-section design for bending:
  The cross-section of the shear key has to be determined from the bending demand by assuming linear strain compatibility in the cross-section

with first yielding of the longitudinal reinforcement. The required section moment of inertia may be assessed with:

$$I_a \geq \frac{M_{Ed,k}\left(d - c_k\right)}{10 f_{ysd}}$$

in which
   $c_k = \lambda_c c_c$ is the depth of the compression zone in the composite region.
   The depth of the composite compression zone has to be assessed iteratively as follows:
   A value of $\lambda_c \geq 1.00$ has to be imposed in order to satisfy the equation:

$$c_k = \frac{f_{ysd}\left(A_s + A_{aft}\left\langle\dfrac{d_{aft} - c_{k,0}}{d - c_{k,0}}\right\rangle + A_{aw}\left\langle\dfrac{d_{aw} - c_{k,0}}{d - c_{k,0}}\right\rangle + A_{afb}\left\langle\dfrac{d_{afb} - c_{k,0}}{d - c_{k,0}}\right\rangle\right)}{\lambda f_{cd} b}$$

where $\langle x \rangle : \left(< 0 = 0; \geq 0 = x\right)$
   Assuming that the shear key is positioned in the geometrical centre of the cross-section with

$$d_a = d_{aw} = h / 2 = 500\,mm$$

Considering the geometrical characteristics of European standard HEB500 profile:

$$b_a = 300\,mm \quad h_a = 500\,mm \quad t_f = 28\,mm \quad t_w = 14.5\,mm$$

This results in the centroids of the shear key components being $d_{aft} = d_a + h_a / 2 - t_f / 2 = 736\,mm$

$$d_{afb} = d_a - h_a / 2 + t_f / 2 = 264\,mm$$

$$d_{aw} = d_a = 500\,mm$$

The corresponding reinforcing areas for the top flange, web and bottom flange are:

$$A_{aft} = A_{afb} = b_a \times t_f = 8400\,mm^2$$

$$A_{aw} = A_a - \left(A_{aft} + A_{afb}\right) = 7060\,mm^2$$

Following the iterative procedure, it gives:
   $c_k = 1.99 \times 175\,mm = 349\,mm$ with $\lambda_c = 1.99$

Hence, the required section moment of inertia of the shear key is

$$I_a \geq \frac{M_{Ed,k}\left(d - c_k\right)}{10 f_{ysd}} = \frac{3307\, kNm\left(928\, mm - 349\, mm\right)}{10 \times 435 N\,/\,mm^2} = 44030 \times 10^4 mm^4$$

The section moment of inertial of the HEB500 profile is:

$$I_a = 107200 \times 10^4 mm^4$$

Embedment length of the shear key:
   The embedment length of the shear key is the maximum between four values. The first, due to strut crushing, is given by the following:

$$l_{v,1} = \frac{V_{Ed.k}}{0.5 v\left(b_a - t_w\right) f_{cd}} = \frac{1565\, kN}{0.5 \cdot 0.5 \cdot \left(300\, mm - 14.5\, mm\right) \cdot 33.3\, MPa} = 329\, mm$$

The length of the shear key is assessed from the condition of yielding of the transverse reinforcement. It is a function of the amount of transverse reinforcement and spacing in the interface region. The definition of the embedment length due to this condition has to be iteratively updated as a function of the verifications for shear.
   The iterative procedure leads to an amount of transverse reinforcement consisting of overlapped stirrups consisting of four legs of 12 mm diameter bars with $A_{sw} = n_{bw} \times A_{dbw} = 4 \times 113\, mm^2 = 452\, mm^2$ spaced at $s_w = 100\, mm$

$$l_{v,2} = \frac{V_{Ed,k} s_w}{A_{sw} f_{ywd}} = \frac{1565\, kN \times 100\, mm}{452\, mm^2 \times 435\, MPa} = 796\, mm$$

A third condition considers that the embedment length should not be less than the projection of the critical strut. Assuming a strut inclination angle of 45 degrees, the embedment length is given by the effective shear depth d_0.

$$l_{v,3} = d_0 \cot \theta = 650\, mm$$

$$d_0 = d - d_a + h_a/2 - t_f = 650\, mm$$

$$l_v = \max\left(l_{v,1}, l_{v,2}, l_{v,3}, h_a\right) = \max\left(329\, mm, 796\, mm, 650\, mm, 500\, mm\right) \approx 800\, mm$$

Shear design: composite region
   Assessment of the shear strength of the composite region:

$$V_{pl,a,Rd} = \frac{A_{a.v}\left(f_{yak}/\sqrt{3}\right)}{\gamma_{M0}} = \frac{8982\, mm^2 \left(460\, MPa/\sqrt{3}\right)}{1.0} = 2385\, kN$$

$\dfrac{V_{Ed,k}}{V_{pl,a,Rd}} = 0.66$, hence the yield strength for assessment of bending strength should be reduced using the ρ parameter:

$$\rho = \left(\dfrac{2V_{Ed,k}}{V_{pl,a,Rd}} - 1 \right)^2 = 0.10$$

In the conceptual design stage, the transverse reinforcement consisted of four legs of 12 mm diameter bars with $A_{sw} = 4 \times 12\,mm = 452\,mm^2$, hence the contribution of the shear reinforcement is given by:

$$V_{Rd,s,k} = A_{sw} z_k f_{ywd} \cot\theta / s_w = 452\,mm^2 \cdot 754\,mm \cdot 435\,MPa \cdot 1/100\,mm$$

$$= 1482\,kN$$

This should be lower or equal to the strut-crushing capacity of the concrete in the panels.

$$V_{Rd,max,k} = 0.5 v f_{cd} (b - b_a) z_k = 0.5 \cdot 0.5 \cdot 33.3\,MPa (700\,mm - 300\,mm) 754\,mm$$

$$= 2511\,kN$$

The shear resistance of cross-sections in the composite region is:

$$V_{Rd,k} = V_{pl,a,Rd} + V_{Rd,s,k} = 3867\,kN$$

$$V_{Rd,k} = 3867\,kN > V_{Ed,k} = 1565\,kN$$

To assess the bending capacity using the force distribution from Section 6.3.4, the following procedure is followed:

$$M_{a,Ed} = \dfrac{V_{Ed,k}}{3} \left(l_v + \dfrac{b_a^2}{8 l_v} \right) = 438\,kNm$$

The elastic moment capacity of the shear key assuming a reduced yield strength due to high shear acting on the steel insert is:

$$M_{a,el,Rd} = W_{el,a} \dfrac{f_{yak}(1 - \rho)}{\gamma_{M0}} = 4287 \cdot 10^3\,mm^3 \dfrac{460\,MPa(1 - 0.1)}{1.0} = 1779\,kNm$$

$$M_{a,el,Rd} = 1779\,kNm > M_{a,Ed} = 438\,kNm$$

Shear design: transition region

$$V_{Rdc,k} = \left(0.6/\gamma_c\right)\left(100\rho_{tot}f_{ck}\,d_{dg}/d_0\right)^{1/3} b d_0$$

$$= \left(0.6/1.5\right)\cdot\left(100\cdot0.0126\cdot50\cdot36/650\right)^{1/3} 700\times650$$

$$= 276kN < V_{Ed,kc} = 1438kN$$

$$\rho_{tot} = \rho_l + \left(0.2l_v/d_0\right)^3 \rho_a = 0.0116 + 0.0149\times0.0682 = 0.0126$$

Transverse reinforcement must be provided. For consistency and based on the detailing rules described above, the same amount of reinforcement provided in the composite region is considered for the transition region.

$$V_{Rd,s,kc} = A_{sw}z_{kc}f_{ywd}\cot\theta/s_w = 452\,mm^2\cdot797\,mm\cdot435\,MPa\cdot1/100\,mm$$

$$= 1567\,N$$

$$V_{Rd,s,kc} = 1567\,kN > V_{Ed,kc} = 1438kN$$

This should be lower or equal to the strut-crushing capacity of the concrete in the panels.

$$V_{Rd,max,kc} = 0.5\nu f_{cd}\left(b_a - t_w\right)d_0 = 0.5\cdot0.5\cdot33.3\,MPa\left(300\,mm - 14.5\,mm\right)650\,mm$$

$$= 1546\,kN$$

$$c_{kc} = \frac{c_c + c_k}{2} = \frac{175\,mm + 349\,mm}{2} = 262\,mm$$

$$z_{kc} = d - \frac{c_{kc}}{2} = 797\,mm$$

$$V_{Rd,max,kc} = 1546\,kN > V_{Ed,kc} = 1438kN$$

Bending verification: composite region

The bending capacity of the composite section assessed using the principles described above is $M_{Rd,k}$=4101 kNm, which is higher than the design bending moment at the column face $M_{Ed,k}$=3307 kNm.

$$M_{Rd,k} = f_{ysd}\left\{A_{sl}\left(d-\frac{c_k}{2}\right) + \Sigma\left[A_{vij}\left\langle\left(d_{vij}-c_k\right)\left(d_{vij}-c_k/2\right)\right\rangle\right]/\left(d-c_k\right)\right\} = 4101\,kNm$$

$$M_{Rd,k} = 4101\,kNm > M_{Ed,k} = 3307\,kNm$$

Bending verification: transition region

The bending capacity of the reinforced concrete cross-section within the transition region M_{Rdc}=2754 kNm is greater than the demand at the composite-to-reinforced concrete interface $M_{Ed,kc}$=2451 kNm.

$$M_{Rd,c} = f_{ysd}A_{sl}\left(d - \frac{c_c}{2}\right) = 2754\,kNm$$

$$M_{Rd,c} = 2754\,kNm > M_{Ed,kc} = 2451\,kNm$$

Design of shear key for full composite action

$$V_{Ed,L,k} = \lambda f_{cd}c_{k,pl}\left(l_v / 8 + b_a\right) = 0.8 \cdot 33.3 \cdot 322\left(800/8 - 300\right) = 3439\,kN$$

$$P_{Rd} = \min\left(0.8f_u\pi d_{ba}^2 / 4; 0.29\alpha d_{ba}^2\sqrt{f_{ck}E_c}\right)/\gamma_a$$

$$= \min\left(0.8 \cdot 450\,MPa \cdot \pi\left(19mm\right)^2 / 4; 0.29 \cdot 0.825 \cdot \left(19mm\right)^2\sqrt{50 \cdot 35654}\,MPa\right)/1.25$$

$$= \min\left(81.6; 125\right)kN = 81.6\,kN$$

$$n \ge \frac{V_{Ed,l}}{P_{Rd}} = \frac{3439}{81.6} = 42.1 \cong 45$$

Considering that the minimum acceptable spacing between shear connectors is 70 mm, the full composite behaviour cannot be achieved with a HEB500 shear key of 800 mm long. Hence, the length of the shear key needs to be extended to 1120 mm to accommodate three rows of 12 mm welded studs, spaced at 70 mm.

A summary of the design procedure and results is presented in Table 6.1, whilst a schematic representation of the designed connection is shown in Figure 6.14.

6.4 CONNECTIONS OF FLAT SLABS TO STEEL COLUMNS

In hybrid flat slab connections, the composite regions are very unlikely to be critical in design. The concrete inside the shear keys is under a form of bi-axial or tri-axial confinement with peaks in the vicinity of the column. As for conventional RC flat slabs, shear-head systems behave as enlarged columns, translating the bending and punching shear critical sections outside

Table 6.1 Results of the design example for the hybrid connection to RC beam

Verification	Region	Cross-section	Capacity $(R_{d,i})$		Demand $(E_{d,i})$	$(R_{d,i}/E_{d,i})$
Bending	Composite	Composite	4188	>	3307	1.27
Shear-steel profile and transverse reinforcement	Composite	Composite	3867	>	1565	2.47
Shear strut crushing	Composite	Composite	2511	>	1565	1.60
Bending	Composite	Shear key	1779	>	438	4.06
Shear	Composite	Shear key	2385	>	1565	1.52
Bending	Transition	Reinforced concrete	2754	>	1931	1.43
Shear-transverse reinforcement	Transition	Reinforced concrete	1567	>	1438	1.09
Shear strut crushing	Transition	Reinforced concrete	1546	>	1438	1.08

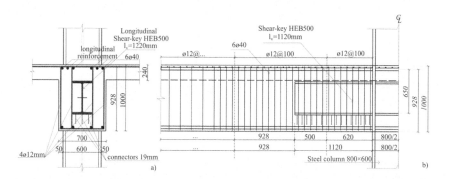

Figure 6.14 Design hybrid connection: a) cross-section, b) side section

the shear-head region. Although some shear key top flange yielding is likely to occur at ultimate, in practical configurations, the weak sections are typically at the composite-to-RC interface.

In the transition zone, the fundamental behaviour of hybrid connections in two-way components is essentially the same, noting that due to flat slab geometry, the strut development is not restricted within a confined space as for beam components but rather able to develop in a fan-shaped manner. As in conventional RC flat slabs, in hybrid systems, punching shear failures are instantaneous and characterized by a dislocation of a conical surface from the flat slab, hence must be avoided. Shear reinforcement can be used to enhance the punching shear strength and ductility of hybrid flat slab connections, noting that the exact contribution of the transverse reinforcement depends on the stud layout, diameter and spacing, shear key system layout and geometry, and slab geometry, among other factors.

As in the case of hybrid connections with one-way components, it is assumed that: (i) plane sections remain plane, and the compatibility relationship between tension and compression is represented by a linear distribution, (ii) the strain in the longitudinal reinforcement bars is the same as that in the surrounding concrete, (iii) the strain in the shear key is a function of the strain in the reinforcement by considering linear cross-sectional strain compatibility, (iv) concrete tensile strength is ignored, (v) the presence of a steel beam in a reinforced concrete component increases locally the strength, stiffness as a function of its embedment length and section size, and influences the component rotational response, (vi) the behaviour in the shear key region is similar to a fully encased composite component.

As shown in Figure 6.5a, the hybrid connection is divided into two in-plane zone types referred to as sectors in which the corresponding equilibrium, constitutive and compatibility laws apply. The sectors that include the shear heads are referred to as "hybrid slab sectors" (i.e. orthogonal to column sides). Each hybrid sector is divided into a composite region containing the shear heads (from the column face to the vicinity of the shear-head tips), a transition region that includes the composite-to-RC interface and a non-composite region outside of the shear-head length. This is schematically illustrated in Figure 6.5b. The sectors without shear heads (diagonals of the component) are referred to as "reinforced concrete (RC) sectors". The design of the composite regions is not covered herein and should follow Eurocode 2 procedures.

6.4.1 Flexural strength

The bending resistance must be verified in all connection regions, such that the design moment should be less than the capacity of the corresponding cross-section. The moment capacity of each characteristic cross-section should be calculated by accounting for linear cross-sectional compatibility and considering that the flexural reinforcement reaches yield. The forces in the shear key in the hybrid sector can be computed by accounting for separate equivalent reinforcement amounts for the top flange, web and bottom flange, and their contribution to the moment capacity should be calculated as a function of their position in the cross-section (Figure 6.15). The equations below can be used to determine the moment resistance of the composite and reinforced concrete connections, respectively.

As noted above, a full shear connection is unlikely to be achieved in the composite region, due to the steel insert not being continuous along the whole span as well as due to the interaction being limited to the steel-concrete bond. Assuming that the shear key cross-section depth is around

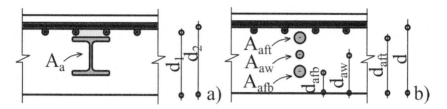

Figure 6.15 Assumed sectional response in composite sections of hybrid connections to flat slabs: a) real case, b) idealized representation

half of the total slab thickness, the compression concrete region would be relatively thin to allow for the insertion of welded shear studs.

The flexural strength V_{flex} of the entire connection can be assessed using limit analysis theorems by accounting for the minimum obtained from several yield line mechanisms. Using this approach, a refined model, employing a hybrid axisymmetric yield line mechanism, was developed by authors Bompa and Elghazouli (2016). Besides the sectorial division of the connection described above, the model assumed that the relatively stiff shear heads, in conjunction with the continuity reinforcement, transfer the entire load from the steel assemblage to the reinforced concrete flat slab component.

For practical application, the refined model is simplified to a closed-form representation of V_{flex}, as shown below. The factor μ_p depends on the type of the connection location ($\mu_p=8$ for interior connections, $\mu_p=4$ for edge connections, and $\mu_p=2$ for corner connections), m_{Rdc} is the bending resistance of the reinforced concrete cross-section, m_{Rdk} is the bending resistance of the composite cross-section, b_a is the width of a shear key, b_{col} is the column width, l_v is the shear key embedment length, and r_s is the distance between the column centre and the zero bending point. The latter can be assumed to be $r_s{\approx}0.22\ L_s$ for the elastic case (L_s is the flat slab span) or can be determined from an elastic analysis. Note that the axisymmetric hybrid model was validated only for interior connections, hence application of the relationship for edge and interior connections ($\mu_p=4$ and $\mu_p=2$, respectively) may require further detailed investigations.

In each sector type, the corresponding constitutive, compatibility and equilibrium relationships apply. The depth of the composite compression zone c_k and of the reinforced concrete compression zone c_c are assessed independently. The notation $<x>$ denotes a Macaulay bracket, indicating that if the value within the brackets is negative, it is considered zero, whilst if the term is positive, the assessed value is considered. The parameter λ defines characteristics of stress distribution in the compression zone. This can be taken as $\lambda=0.8$ for normal concrete ($f_{ck}{\leq}50\ MPa$). Other methods to assess the cross-sectional capacity can be used, as long as they follow the assumptions and principles described before. The assessment of the depth of the composite compression zone is carried out by obeying linear strain compatibility, equilibrium and constitutive conditions within the cross-section. This can be achieved by using an iterative procedure in which the initial depth of the compression zone $c_{k,0}=\lambda_c c_c$ and $\lambda_c \geq 1.0$, so that, eventually, equilibrium is satisfied and $c_k=c_{k,0}$.

$$V_{flex} = \mu_p m_{Rd,c}\left[1+4\frac{b_a}{b_{col}}\frac{l_v}{r_s}\left(\frac{m_{Rd,k}}{m_{Rd,c}}-1\right)\right] \tag{6.28}$$

$$m_{Rd,k} = f_{ysd}\left\{\rho_l d\left(d-\frac{c_k}{2}\right)+\Sigma\left[\rho_{vij}d_{vij}\left\langle(d_{vij}-c_k)(d_{vij}-c_k/2)\right\rangle\right]/(d-c_k)\right\} \tag{6.29}$$

$$c_k = \frac{f_{ysd}\left(\rho_l d+\rho_{aft}d_{aft}\left\langle\dfrac{d_{aft}-c_{k,0}}{d-c_{k,0}}\right\rangle+\rho_{aw}d_{aw}\left\langle\dfrac{d_{aw}-c_{k,0}}{d-c_{k,0}}\right\rangle+\rho_{afb}d_{afb}\left\langle\dfrac{d_{afb}-c_{k,0}}{d-c_{k,0}}\right\rangle\right)}{\lambda f_{cd}} \tag{6.30}$$

$$m_{Rd,c} = \rho_l d f_{ysd}\left(d-\frac{c_c}{2}\right) \tag{6.31}$$

$$c_c = \frac{f_{ysd}A_s}{\lambda f_{cd}b_c} \tag{6.32}$$

$$\text{and} \quad \rho_{vij} = A_{vij}/(b_{sr0}d_{vij}) \tag{6.33}$$

and b_{sr0} is the unit moment strip
   where

$$\langle x\rangle:(<0=0;\geq0=x)$$

The tension longitudinal reinforcement in the composite region of the hybrid connection should be greater than 0.4% but less than 2.0%, assessed by considering the bare concrete section. The longitudinal reinforcement should be continuous above the column support. This can be achieved by drilling through the column flanges or by passing the rebars around the column allowing for adequate spacing such that the concrete can be placed and compacted satisfactorily for the development of adequate bond. To avoid congestion of reinforcement and localized stress concentrations, lapping of bars, welding and mechanical devices should not be used within the composite region and at the interface. Laps, welding and mechanical couplers are allowed only in the non-composite region in accordance with Eurocode 2 for reinforced concrete components.

6.4.2 Punching shear strength

6.4.2.1 Punching shear resistance of slabs without shear reinforcement

This section deals with punching shear strength design reinforced concrete flat slab components provided with cruciform shear heads at the hybrid connection to interior rectangular steel columns. Application of the equations described below to edge and corner connections should be made by analogy, considering a reduced control perimeter and corresponding parameters as the main behavioural characteristics captured in the design model. However, this would need to be supported by more detailed investigations.

The key design parameter is the embedment length l_v of the shear head that dictates the location of the critical shear section and provides a definition of the critical perimeter. The design expressions described below refer to the punching shear strength of hybrid connections without moment actions provided with or without transverse reinforcement. As illustrated in Figure 6.16, the method considers that the shear heads are relatively stiff and translate the critical shear region at the composite to the concrete interface. The force transfer is achieved by struts supported on the bottom flanges of the four shear keys, hence force-transferring struts should be provided with adequate support width.

To ensure a reliable shear-head stiffness, the depth of the shear key cross-section, using common hot rolled class 1 and 2 steel sections, should be in the range of 50–70% of the effective bending flat slab depth ($h_a = 0.5–0.7d$). Additionally, the embedment length should be greater than twice the shear key depth ($l_v \geq 2\, h_a$). The length should be limited to five times its depth, since above this size ($l_v \leq 5\, h_a$), the shear key would behave as a relatively flexible insert and would have limited stiffness to ensure adequate force transfer. In such cases, flexural shear key yielding governs the response at a capacity lower than the punching shear strength.

For the calculation of the punching shear resistance, a shear-resisting control perimeter b_0 is used. The location of the critical section is dependent on the shear key length-to-slab radius ratio l_v/r_s and influenced by its shape and section size. A detailed analysis of stress fields described elsewhere (Bompa and Elghazouli, 2016) showed that b_0, required for punching shear assessments in the case of hybrid connections provided with cruciform shear heads, can be expressed in terms of a critical length l_0, as shown in Figure 6.16. The critical length defining the critical section can be safely assumed as located at half the effective shear depth d_0 from the shear key tip. As shown in Figure 6.16, b_0 is defined for each shear head by an arc length with a radius equal to the in-plane strut projection d_0 plus two

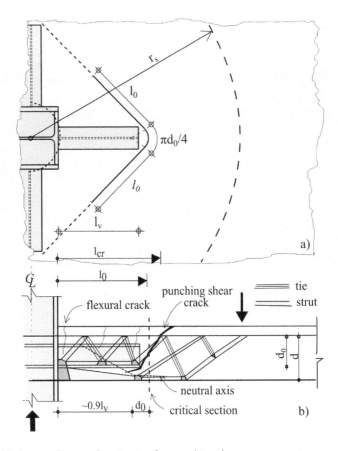

Figure 6.16 Assumed control perimeter for punching shear assessments

critical lengths l_0. For short shear heads, the diagonal lines of the control perimeter extending from the shear-head tip regions could join, resulting in a closed shape. It can be evaluated using the equations below, as the minimum of the above two conditions. For a general case (e.g. non-rectangular columns), the critical perimeter $b_0 = V_{Ed}/v_{perp,max}$ can be assessed by means of a shear field analysis using Model Code 2010 (fib, 2012) recommendations where the maximum perpendicular shear force per unit width $v_{perp,max}$ can be assessed by means of elastic analysis.

$$b_0 = \min(\pi d_0 + 8 l_0, \pi d_0 + 4\left[l_0 + (b_{col} - b_a)/2\right]\sqrt{2}) \tag{6.34}$$

$$l_0 = l_v + d_0/2 \geq 2 b_a \tag{6.35}$$

$$d_0 = d - d_{afb} - t_f/2 \tag{6.36}$$

Previous experimental investigations indicated a high dependency between the punching crack pattern on the top face of the slab and the embedded length of the shear head (Bompa and Elghazouli, 2016). The ratio between the embedment length l_v and the radial crack length l_{cr}, represented by the in-plane distance between the column face and the punching shear crack at the intersection with the flexural reinforcement, as shown in Figure 6.16, is in the range of l_{cr}/l_v=1.00–2.04 (average of 1.38) in the hybrid sectors and l_{cr}/l_v=0.74–1.90 (average of 1.30) in the RC sectors. This corresponds to an average distance of $3.12d$ from the column face in the hybrid sectors, and $2.80d$ in the RC sectors – which is higher than typically seen in RC flat slabs (1.0–2.0d). This indicates that, because of the presence of the shear head, the failure surface appears outside the shear-head region. Although in tests the critical section was not always at $d_0/2$ due to loading arrangement and direct load transfer to supports, in continuous flat slab components, the crack inclination would follow the elastic compression stress field, which is at 45° to the slab.

As mentioned above, in hybrid flat slab connections, the critical section in shear would be within the transition region, around the composite-to-RC interface. The connection design must ensure adequate punching shear strength V_{Rdc}, greater than the design shear force V_{Ed}, within this region. The method described here adopts the fundamentals of the revised Eurocode 2 draft which is based on the principles of the Critical Shear Crack Theory (Muttoni, 2008). The closed-form equations below are based on a refined model developed previously by authors Bompa and Elghazouli (2016).

To capture the influence of the shear-head embedment length and section size on the punching shear capacity of the connection, the parameter $k_{pb,kc}$ incorporates factors specific to the hybrid configuration. This is used to assess the punching shear resistance $V_{Rd,kc}$ of hybrid flat slab components not requiring shear reinforcement in the transition region, where shear critical sections are expected to develop. The resistance $V_{Rd,kc}$ is a function of the flexural reinforcement ratio of the reinforced concrete section ρ_l, concrete characteristic compressive strength f_{ck}, a size effect parameter d_{dg} that is the sum of the reference aggregate size d_{g0}=16 mm and the maximum aggregate size d_g in the concrete, the control perimeter b_0, the effective shear depth d_0, and the $k_{pb,kc}$ parameter mentioned before. The terms on the right-hand side under the square root represent the shear-gradient enhancement factor, whilst the terms in the square brackets capture the influence of the composite region on the punching shear strength. Note that the equations below were validated only for interior connections (μ_p=8), hence their application for edge and interior connections (μ_p=4 and μ_p=2, respectively) may require further detailed investigations.

$$V_{Rd,kc} = \frac{0.6}{\gamma_c} k_{pb,kc} \left(100\rho_l f_{ck} \frac{d_{dg}}{r_s} \right)^{\frac{1}{3}} b_0 d_0 \leq \frac{0.6}{\gamma_c} \sqrt{f_{ck}} b_0 d_0 \qquad (6.37)$$

$$k_{pb,kc} = \sqrt{35\mu_p \frac{d}{b_0}\left[1+4\frac{b_a}{b_c}\frac{l_v}{r_s}\left(\frac{m_{Rk}}{m_{Rc}}-1\right)\right]}$$
(6.38)

$$\mu_p = \begin{vmatrix} 8 \to \text{int } erior \\ 4 \to edge \\ 2 \to corner \end{vmatrix}$$
(6.39)

6.4.2.2 Punching shear resistance of slabs with shear reinforcement

In the case of hybrid connections provided with shear reinforcement, the ultimate limit state is expected to be similar to that in conventional RC components (Figure 6.17). Experimental observations indicated that stud shear reinforcement added to hybrid components with shear heads enhances the strength and ductility. The increase in ductility is primarily attributed to the post-elastic response of the flexural bars, whereas the strength increase results from the activation (i.e. yield and potential fracture) of a number of transverse bars intersected by the critical crack, with failure occurring due to punching shear within the shear-reinforced region, with the shear head remaining largely elastic.

In conventional RC flat slabs, the presence of transverse reinforcement can increase the ultimate punching shear strength. In this case, possible failure modes include: (i) failure within the shear-reinforced region ($V_{Rd,cs,in}$) in which the critical crack intersects transverse bars, (ii) failure outside the shear-reinforced region ($V_{Rd,cs,out}$) in which the critical surface does not cross the shear reinforcement, and (iii) due to strut crushing (located between the column face and first reinforcement perimeter – $V_{Rd,max}$).

The design method described below assumes that the failure modes in hybrid flat slab connections provided with cruciform shear heads and provided with headed shear studs are the same as in conventional RC flat slabs. Note that similarly to hybrid connections without transverse reinforcement, the critical section is located around the composite-to-RC interface, within the transition region. Considering the three potential failure modes, the punching shear strength of hybrid connections provided with headed shear studs is the minimum of the three conditions $V_{R,cs}=min(V_{Rd,cs,in}, V_{Rd,max}, V_{Rd,cs,out})$. Flexural failure could govern if yielding occurs in the longitudinal reinforcement prior to yielding of the transverse bars.

The capacity of a hybrid connection provided with stirrups that fails within the shear-reinforced region is based on the cumulative contribution of the concrete $V_{Rd,c}$ and the transverse bars $V_{Rd,s}$, reduced by η_c and η_s, respectively. The parameter η_c captures the concrete contribution at critical crack development, whilst η_s accounts for the level of activation of the transverse bars. The latter depends on the slab rotation, captured by the

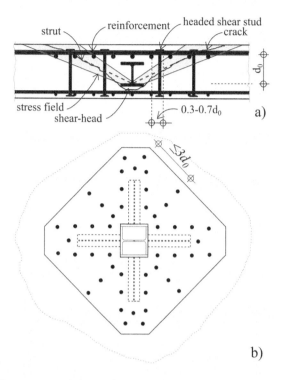

Figure 6.17 a) Cross-sectional detail of a hybrid flat slab provided with headed studs, b) Suggested headed shear studs layout

slenderness ratio (r_s/d) as well as other geometry parameters. In conventional RC flat slabs, the shear reinforcement A_{sw} depends on the radial and tangential spacing of the transverse bars located at the control perimeters. In hybrid connections, A_{sw} is dependent on the embedment length and width of the shear head. The amount of reinforcement A_{sw} accounted for in a design is the first row of well-anchored transverse bars situated in the vicinity of the edges of the shear head (Figure 6.17)

To ensure reinforcement activation, the headed studs should be located in the region of $0.3–0.7d_0$ from the flange edges considering that the shear cracks generally develop under governing struts that originate from the bottom flanges of the steel profile at inclination angles of about 45°. Additionally, in order to allow a smooth force transfer, the bottom flange of the shear heads should be relatively stiff and at least $b_a > d/2$ wide based on test ranges considered in previous research investigations.

$$V_{Rd,cs,in} = \eta_c V_{Rd,c} + \eta_s V_{Rd,s} \qquad (6.40)$$

$$V_{Rd,s} = A_{sw} f_{ywd} \qquad (6.41)$$

$$\eta_c = V_{Rd,c}/V_{Ed} \tag{6.42}$$

$$\eta_s = \left(\frac{r_s}{d}\right)^{1/2}\left(\frac{0.8}{\eta_c\sqrt{8\mu_p d/b_0}}\right)^{3/2} \leq 0.8 \tag{6.43}$$

As pointed out before, the headed shear studs provided strength and ductility enhancements of tested hybrid flat slab connections (Bompa and Elghazouli, 2016). Close examination of the test results suggested that in the case of components with double-headed shear studs, the punching shear crack was initiated in a similar manner to hybrid slabs without transverse reinforcement. Failure was initiated in the hybrid sectors at the composite-to-RC interface and further propagated towards the conventional RC sectors (i.e. diagonals of the slab) whilst activating several transverse bars. Based on the observations of detailed analysis on the strain gauges as well as crack pattern and kinematics, it was shown that the number of studs activated by the punching shear crack was between 24 and 28 for the two configurations tested.

The exact amount of activated transverse bars is influenced by the shear flow within the slab which, from a mechanical point of view, is dependent mainly on the strut support and, from a practical point of view, on the shape of the shear head. For the test components under discussion, the headed shear studs located at the composite-to-RC interface were either activated last or provided no contribution. Therefore, in the design of components with straight-cut shear heads and with shear reinforcement, the transverse bars located at the shear-head tip should not be taken into consideration. However, they must be provided in the flat slab to avoid failures outside of the shear-reinforced region. The comments above highlight the need for further investigations on the response of hybrid connections with shear heads, and the practical design of such connections should be treated with careful attention.

6.4.3 Shear-head properties

The embedment length l_v of a shear head may be determined from the assumption that the critical section is situated around the composite-to-concrete interface l_0, which results from the length of a required control perimeter b_0 as a function of the design shear force V_{Ed}. The assessment of the required control perimeter b_0 should follow an iterative procedure, being updated with final material and geometric parameters. In the first iteration, $k_{p,kc}$ can be assumed equal to 4.0. Consequently, the critical length l_0 and the required embedment length l_v can be assessed using the equations below.

Non-linear numerical assessments showed that relatively long shear heads, with relatively small depth h_a compared to slab thickness, are ineffective, as web shear failure may govern ahead of ensuring reliable force transfer to the steel column (Bompa and Elghazouli, 2017). On the other hand, short shear heads, in the range of $l_v/h_a{\leq}1.0$, are unable to support the force-transferring struts, which may lead to compression yielding of the bottom flange and slip. For components with intermediate l_v and intermediate shear-head depths $(h_a/d{\geq}0.43)$, controlled failures are generally obtained, and the design approach presented herein captures well the main behaviour. It is worth noting that, depending on the flexural reinforcement ratio ρ_l, at least one of the steel tension components may yield, with a failure mode similar to flexural punching that develops in conventional RC flat slabs.

$$b_0 \geq \frac{V_{Ed}}{\dfrac{0.6}{\gamma_c} k_{pb,kc} \left(100\rho_l f_{ck} \dfrac{d_{dg}}{r_s}\right)^{\frac{1}{3}} d_0} \tag{6.44}$$

$$l_0 \geq \max\left[\left(b_0 - \pi d_0\right)/8; \left(b_0 - \pi d - 2\left(b_{col} - b_a\right)\sqrt{2}\right)/\left(4\sqrt{2}\right)\right] \tag{6.45}$$

$$l_v \geq l_0 - d_0/2 \geq \max(2h_a, d/2) \tag{6.46}$$

The shear-head section size can be determined by means of linear strain compatibility in the composite cross-section by assuming that each shear key component is represented by a single reinforcement bar located at its geometrical centre. The stress in a reinforcement component and its contribution to the moment resistance is a function of the cracked stiffness $I_{k,cr}$ and geometry of the component. The cracked stiffness depends on the geometrical characteristics of the constituent elements of the cross-section. For RC components, the cracked moment of inertia at ultimate varies between 30–70% of the un-cracked cross-section. Examination of tests and numerical models of hybrid connections pointed to an average ratio of $c_k/d=0.37$ at ultimate, leading to a ratio of the shear-head moment of inertia I_a to the cracked moment of inertia of the entire cross-section per unit width of $I_a/I_{k,cr}=0.25$.

Moreover, the shear key cross-section should be determined from the highest design moment $M_{Ed,k}$ and shear force $V_{Ed,k}$ within the composite region. Considering linear strain compatibility within the composite cross-section with yield occurring in the flexural reinforcement, the shear key second moment of inertia can be determined as a function of the design moment m_{Ed}, slab effective bending depth d, and yield strength of the flexural reinforcement f_{ysd}. Note that $\mu_p=8$, $\mu_p=4$ and $\mu_p=2$, can be assumed for interior, edge and corner connections, respectively, yet application for edge and corner cases should be supported by supplementary detailed investigations.

In addition to the shear key design using the design moment, this needs to be provided with sufficient shear area $A_{a,v}$ in order to avoid web failures of the steel profile. The required shear area is a function of the design shear V_{Ed}, the number of shear keys in the shear head ($n_v=4$ for cruciform shear heads) and the yield strength of the material in the steel profile f_{yad}.

Considering that the force is transferred from the slab to the column through struts supported on the bottom flange, the width b_a of the shear key dictates the cross-sectional strut thickness and, consequently, the amount of force transferred. In order to ensure a smooth transfer, the bottom flange should be relatively stiff and have adequate width to avoid failure in compression in the steel insert. In addition to the equation below to assess the b_a, the shear key section size should also comply with the $h_a/d \geq 0.5$ limit for standard European sections. For values of $h_a/d < 0.5$, the shear key web area is likely to be insufficient to carry the design shear.

$$I_a \geq \frac{m_{Ed}}{10} \frac{d}{f_{ysd}} \left(1 - \frac{c_k}{d} \right) \tag{6.47}$$

$$A_{av} \geq \frac{5}{4} \frac{V_{Ed}}{n_v} \frac{\sqrt{3}}{f_{yak}} \tag{6.48}$$

$$b_a \geq \frac{V_{Ed}}{n_v} \frac{2}{v f_{cd} l_v} \tag{6.49}$$

Key observations from numerical simulations on hybrid connections to flat slabs showed that force distribution per unit width (v/v_{max}) at the top flange of the shear head as a function of the embedment length l_v and, as for one-way components, can be divided into two regions (Bompa and Elghazouli, 2017). In all cases, the peak value was recorded within a region of $h_a/4$ from the column face. For short and relatively rigid shear heads, the distribution takes nearly a triangular form, whereas for flexible steel inserts, it shows a non-linear form. In a simplified manner, it can be represented by two rectangular regions delimited by $h_a/4$. Close inspection of the numerical results showed that the average reaction forces in each region vary with shear-head flexibility, and the amount of force transferred through each region can be expressed as a function of the distribution factor κ. The moment carried by each shear key $M_{a,Ed}$ can be expressed as a function of the shear action V_{Ed}/n_v and the assumed distribution in Figure 6.18. As above, $n_v=4$ for cruciform shear heads.

$$M_{a,Ed} = \frac{V_{Ed}}{n_v} \left[\kappa \frac{b_a}{8} + (1 - \kappa) \left(\frac{l_v}{2} + \frac{b_a}{8} \right) \right] \tag{6.50}$$

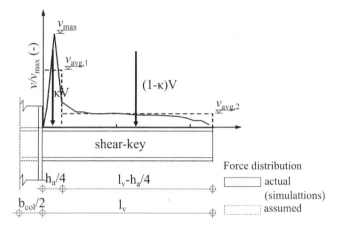

Figure 6.18 Reaction force distribution for shear keys in flat slabs

$$\kappa = \frac{1}{3}\left(1 - \frac{l_v}{r_s}\right) \tag{6.51}$$

As for one-way components, in hybrid connections to flat slabs, the full plastic capacity of the steel insert is generally not reached, since the behaviour is governed by the interaction properties between the shear head and embedding concrete. The numerical results showed that the shear-head/concrete slip is dependent on the contact surface and, consequently, the embedment length l_v. Hence, the shear keys develop inelastic behaviour, primarily at their top flange, at a fraction of their full plastic capacity. This fraction can be estimated by accounting for the k_m factor which was determined from the numerical simulations in which the yielding of the shear-head flange was recorded (Bompa and Elghazouli, 2017). Ultimately, the shear key cross-section needs to be verified for shear.

$$M_{a,Ed} \le k_m W_{a,pl} f_{yad} \tag{6.52}$$

$$k_m = \left(65\rho_l\right)^{1/3}\left(\frac{d_0}{b_a}\right)\left(\frac{l_v}{r_s}\right)^{2/3} \tag{6.53}$$

$$\frac{V_{Ed}}{n_v} \le V_{pl,a,Rd} \tag{6.54}$$

$$V_{pl,a,Rd} = \frac{A_{a,v}\left(f_{yak}/\sqrt{3}\right)}{\gamma_{M0}} \tag{6.55}$$

6.4.4 Validation and practical recommendations

The expressions proposed for the design of hybrid connections between steel columns and reinforced concrete flat slabs by means of shear-head systems were validated against an experimental database consisting of 16 tests carried out by authors and from the literature (Lee et al., 2008; Kim et al., 2014; Eder et al., 2011; Bompa and Elghazouli, 2017). The main parameters varied were as follows: slab thickness h=150–225 mm, concrete compressive strength f_c=17.1–39.2 MPa, flexural reinforcement ratio ρ_l=0.3–1.4 %, shear key embedment length l_v=200–770 mm, shear key depth h_a=51–150 mm, l_v/h_a=2.2–8.0, and steel column size b_{col}=180–500 mm. Two of the tests were provided with transverse reinforcement in the form of double-headed shear studs. To provide detailed insights into the physical behaviour of a wide range of hybrid configurations outside of the existing test database, parametric investigations on 92 models were undertaken (Bompa and Elghazouli, 2017). The non-linear numerical procedures were validated against the tests carried out by the authors on hybrid connections as well as an established test series on interior RC flat slab connections (Guandalini et al., 2009; Bompa and Elghazouli, 2016, 2020). The main parameters varied in the parametric investigations were as follows: shear-head embedment length-to-depth ratio (l_v/h_a=0.5–5.0), reinforcement ratio ρ_l of 0.33–2.20%, slab effective bending depth (d=140–330 mm), slab span, and concrete strength f_c=29–80 MPa.

The equations for assessing the flexural strength have been validated against the test and numerical models which indicated failure by yielding of flexural reinforcement, using an axisymmetric mechanical model which predicts the complete load-rotation response of the hybrid connection. As shown by the comparison from Figure 6.19, the flexural strength of the component, accounting for full yielding of flexural reinforcement, is estimated reasonably well. The ratio between the reported strength and that predicted is V_{test}/V_{flex}=0.96. The analytical results showed consistency with test results since the full flexural capacity was not reached during the tests, yet yielding in the longitudinal reinforcement was recorded.

As noted above, the punching shear strength expressions were developed using the fundamentals of an established model (Muttoni, 2008; Muttoni et al., 2018), which requires the full load-rotation response and a failure criterion. The punching shear capacity is obtained by the intersection between the two curves. For hybrid connections, the axisymmetric model mentioned above as well as a simplified bi-linear model developed by authors enable the assessment of the full deformational response (Bompa and Elghazouli, 2016). The failure criterion follows the original RC representation, noting that the failure surface and the control perimeter are shear-head dependent. Close inspection of analytical models for conventional RC connections indicated that these cannot be used to predict the rotational response or the flexural strength. On the other hand, the proposed methods showed good

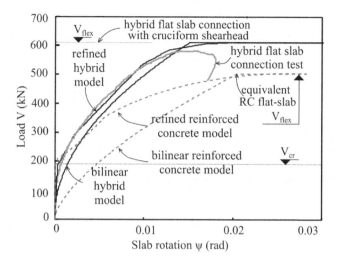

Figure 6.19 Validation of the flexural strength response of interior hybrid connections to flat slabs

agreement with the test results, as described in detail elsewhere (Bompa and Elghazouli, 2016, 2017).

For practical application, the proposed models, which require a more complex iterative procedure and assessment of structural parameters, are less used in common design, hence the expressions developed to assess the slab rotation ψ and the influence of the shear head on the stiffness of the connection have been incorporated in a closed-from in the $k_{pb,kc}$ parameter. This code-like format approach, employed in the revised Eurocode 2, ensures consistency between the formulations proposed for the design of hybrid connections to one-way and two-way components.

The punching shear design equations from Section 6.4.2 are used to assess the strength of the test components and numerical models described above. The punching shear strength was computed by considering assessed material strengths by laboratory testing and material safety factors equal to unity (γ_i=1.0). The predicted results depicted in Figure 6.20 show good agreement with those obtained from tests and simulations for a wide range of geometries with various shear key embedment lengths (l_t/h_a=1.0–4.0). The average $V_{test/num}/V_{calc}$=1.15, the coefficient of variation is 0.12, and the 5% percentile is 0.94. Although the latter is below 1.00, the estimates are acceptable, noting that more conservative estimations could be obtained if design safety factors are used. It is also noteworthy that the predictions for the transversely reinforced hybrid connections, marked with triangles in the figures, are conservative with $V_{test/num}/V_{calc}$=1.14.

In terms of implications on practical design, close examination of the above design indicates that the increase in reinforcement from a low

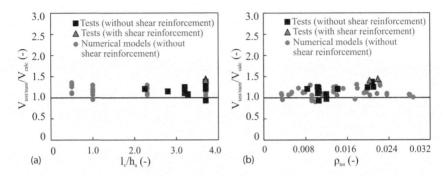

Figure 6.20 Validation of punching shear models for hybrid connections: a) embedment
length, b) combined reinforcement ratio

ratio of $\rho_l=0.3\%$ to a relatively high ratio of $\rho_l=2.0\%$ would lead to
about a five-fold enhancement in stiffness and strength. In contrast, possible increases in shear-head section sizes, for practical ranges of flat
slab thicknesses, would have a comparatively insignificant influence on
the stiffness and flexural strength. It appears therefore that the use of a
high conventional reinforcement ratio in conjunction with small shear-
head section sizes (e.g. $\rho_l=1.1\%$ and HEB100) would be more effective
than low reinforcement ratios combined with larger shear-head section
sizes (e.g. $\rho_l=0.3\%$ and HEB200).

It is generally observed that the low amounts of ρ_l lead to flexural
failures and that high-design shear V_{Ed} requires very long shear heads
with $l_v>r_s$. Components provided with low to intermediate reinforcement (e.g. $\rho_l=0.75\%$) may also reach their flexural strength when small
shear-head section sizes are employed. Typically, irrespective of l_v, intermediate to high ρ_l produces elastic reinforcement behaviour. Generally,
for the design shear force V_{Ed}, low ρ_l ratios require shear heads with
higher l_v/r_s in comparison with intermediate and high ρ_l, which may
necessitate the use of short to intermediate shear heads. In addition to
this, close inspection of numerical results indicated ineffective behaviour for short shear heads ($l_v/r_s<0.1$), since they were not able to ensure
a smooth force transfer through struts. Also, available data showed that
for $0.1<l_v/r_s<0.55$ the behaviour is generally effective, and the lengths
seem practical. These observations point to effective use of shear heads
in design, with embedment length-to-slab radius ratios within the range
of $l_v/r_s=0.2-0.5$, mainly due to their practicality and stable structural
behaviour. Moreover, the shear key embedment length to depth should
be greater or equal to $l_v/h_a\geq2.0$.

Besides the validation of the expressions for bending, punching shear
and embedment length, the results of the numerical simulations resulted
in expressions for the design of the shear-head cross-section. Figure 6.21

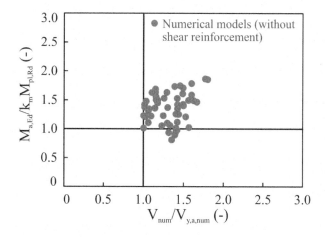

Figure 6.21 Validation of shear key design method for hybrid connections to flat slabs

shows the shear key design moment-to-moment capacity ratio $M_{a,Ed}/k_m M_{pl,Rd}$ against the $V_{u,num}/V_{ya,num}$ ratio, in which $V_{u,num}$ is the ultimate strength obtained from non-linear simulations and $V_{ya,num}$ is the force at which yielding developed in the top flange. Values below 1.0 indicate that no plastic behaviour is predicted $(M_{a,Ed}/(k_m M_{a,pl})<1.0)$, and elastic shear key behaviour was developed in simulations $(V_{u,num}/V_{ya,num}<1.0)$. It is shown that the assumed force distribution (Figure 6.18) in conjunction with the considered moment behaviour reflects good agreement between predictions and numerical results.

Hybrid flat slab connections with fully embedded shear heads can develop a ductile response either by yielding of the flexural reinforcement and/or the shear keys. Yielding of flexural reinforcement would normally occur for low to intermediate flexural reinforcement ratios, whilst yielding of the shear keys would normally occur by having relatively small steel sections or excessively long inserts. Low flexural reinforcement ratios may not be feasible due to design demand or geometry constraints, and small shear key sections or long inserts may not be practical. Whilst some experimental evidence to support the use of fully embedded shear heads in flat slabs requiring ductile response exists (Lee et al., 2019), further investigations seem necessary to assess the complete response in detail and design guidance for practical configurations is hence limited. An alternative to fully embedded shear heads is partially embedded systems in which isolated steel beams act as fuses. In this configuration, the cyclic response is governed by inelasticity at the steel inserts with the reinforced concrete region remaining largely elastic (Eder et al., 2011). Design guidance for hybrid flat slab connections with partially embedded shear heads is available in Eder et al. (2012).

6.4.5 Design example – Connection of steel columns to flat slabs

6.4.5.1 Description of structure

The design of a two-way flat slab floor system is required in which a hybrid structural system made of steel columns and reinforced concrete flat slab is chosen. The force transfer is made through a short steel insert, referred to as the shear key, welded to the steel column and fully integrated into the reinforced concrete floor.

Exposure class XC1 (Concrete inside buildings with low air humidity)

Structural class S4, for C30 -> reduce with class 1, results in S3

The moment spans are $L_x = 6.0\,m$ and $L_y = 5.0\,m$

Column sizes are $b_{c1} = 240\,mm$ and $b_{c2} = 280\,mm$

The design bending moments and shear forces are obtained through an elastic FE analysis. The design forces and moments are extracted at the column face. Their values are:

Shear force: $V_{Ed} = 467\,kN$

Bending moments: $m_{Ed,x} = 70\,kNm/m$ and $m_{Ed,y} = 54\,kNm/m$

Twisting moment: $m_{Ed,xy} = 8.6\,kNm/m$

The zero bending moment section is located at $r_{s,x} = 1.25\,m$ and $r_{s,y} = 1.09\,m$ from column centre.

6.4.5.2 Material properties

Concrete grade C30 $f_{ck} = 30\,MPa$ $\gamma_c = 1.5$ $f_{cd} = 20\,MPa$ $d_g = 20\,mm$

Longitudinal steel grade B500S $f_{ysk} = 500\,MPa$ $\gamma_s = 1.15$ $f_{ysd} = 435\,MPa$

Transverse steel grade B500S $f_{ywk} = 500\,MPa$ $f_{ywd} = 435\,MPa$

Structural steel Grade S355 $f_{yvk} = 355\,MPa$ $\gamma_{M0} = 1.00$ $f_{yvd} = 355\,MPa$

$$E_s = E_{sw} = E_a = 200\,GPa$$

6.4.5.3 Reinforcement design

In the conceptual design stage, the slab depth is determined from the three conditions (i.e. minimum stiffness, fire requirements REI90 and technological aspects – the depth of the shear key should be at least half of the effective bending depth).

$$h_s = \max(h_{s,init,a}, h_{s,init,b}, h_{s,init,c}) = \max(171, 200, 222) \cong 230\,mm$$

$$h_{s,init,a} = L_x / 35 = 6.0m / 35 = 172\,mm \text{ (stiffness)}$$

$$h_{s,init,b} = 200\,mm \text{ (fire)}$$

Assuming that the effective bending depth d of the slab is around 90% of its thickness and that the shear key depth $h_{a,min}$ should be between 50–70% of the effective bending depth,

$$d \cong 0.9h_{s,init,c} \rightarrow 2h_{a,min} \cong 0.9h_{s,init,c} \rightarrow h_{s,init,c} = 222\,mm \qquad \text{(composite)}$$

in which $h_{a,min} \geq d/2$ and $h_{a,min} = h_{a,HEB120} = 120\,mm$

The flat slab thickness is then chosen as $h_s = 230\,mm$

The reinforcement design is simplified in this example due to similar moment spans in both orthogonal directions. An iterative procedure leads to a total tension reinforcement amount for both orthogonal directions consisting of $11\phi12/m$ $(A_{sl} = 1244\,mm^2)$

It is assumed that the tension reinforcement must be able to carry the bending and torsion $m_{Rd,c} \geq m_{Ed,max}$ moment per unit strip

The plastic moment assessed per moment unit strip is:

$$m_{Rd,c} = \rho_l df_{ysd}\left(d - \frac{c_c}{2}\right) = 0.0063 \times 198 \times 435(198 - 34/2) = 98.0\,kNm/m$$

in which the average effective bending depth is:

$$d = (d_x + d_y)/2 = (204 + 192)/2 = 198\,mm$$

with $d_x = h - c_{nom} - d_{blx}/2 = 230 - 20 - 6 = 204\,mm$

$$d_y = h - c_{nom} - d_{blx} - d_{bly}/2 = 230 - 20 - 12 - 6 = 192\,mm$$

$$c_{min} = \max(c_{min,b} = 12\,mm, c_{min,dur} = 10\,mm, 10\,mm) \rightarrow c_{nom} = 20\,mm$$

the flexural average reinforcement ratio

$$\rho_l = \sqrt{\rho_{lx} \times \rho_{ly}} = \sqrt{\frac{1244}{204 \times 1000} \times \frac{1244}{192 \times 1000}} = \sqrt{0.0061 \times 0.0065} = 0.0063$$

$$\rho_{li} = A_{sli}/(b_{sr0}d_i)$$

and the average depth of the compressive zone in the concrete region
$$c_c = \frac{\rho_l df_{ysd}}{\lambda f_{cd}} = \frac{0.0063 \times 198\,mm \times 435\,MPa}{0.8 \times 20\,MPa} = 33.8\,mm$$

The average design bending moment is determined from:

$$m_{Ed,max} = \max(m_{Ed,x,avg}, m_{Ed,y,avg}) = 78.2\,kNm$$

$$m_{Ed,x,avg} = (m_{Ed,x} + |m_{Ed,xy}|)b_{sr,x} = 78.2\,kNm$$

$$m_{Ed,y,avg} = \left(m_{Ed,y} + \left|m_{Ed,xy}\right|\right)b_{sr,y} = 63.6\,kNm$$

in which $b_{sr,x}$ and $b_{sr,y}$ are the orthogonal bending active widths for punching shear assessments (see Model Code 2010)

$$b_{sr,x} = b_s / 2 + b_{c1} / 2 = 1.00\,m$$

$$b_{sr,y} = b_s / 2 + b_{c2} / 2 = 1.02\,m$$

$$b_s = 1.5\sqrt{r_{s,x} \times r_{s,y}} = 1.5\sqrt{1.25 \times 1.09} = 1.75\,m$$

Shear key design:

Design for shear

The required plastic resistance of one shear key, in which n_v is the number of shear keys is: $V_{a,pl,Rd} \geq V_{a,Ed} = V_{Ed} / n_v = 476\,kN / 4 = 119\,kN$

$$A_{a,v} \geq \frac{5}{4}\frac{V_{a,Ed}}{f_{yak} / \sqrt{3}} = \frac{119\,kN}{355\,MPa / \sqrt{3}} = 726\,mm^2$$

HEB120 profile is chosen, although a smaller section would be sufficient. This is to minimize the influence of shear on the moment capacity of the shear key.

$$A_{a,v} = 1096\,mm^2$$

Design for bending

The cross-section of the shear key must be determined from the bending demand by assuming linear strain compatibility in the cross-section with first yielding of the longitudinal reinforcement. The section moment of inertia is obtained from:

$$I_a \geq \frac{m_{Ed,max}\left(d - c_k\right)}{10 f_{ysd}}$$

in which

$c_k = \lambda_c c_c$ is the depth of the compression zone in the composite region.

The depth of the composite compression zone must be assessed iteratively as follows:

A value of $\lambda_c \geq 1.00$ has to be imposed in order to satisfy the equation:

$$c_k = \frac{f_{ysd}\left(\rho_l d + \rho_{vft} d_{vft}\left\langle \frac{d_{vft} - c_{k,0}}{d - c_{k,0}}\right\rangle + \rho_{vw} d_{vw}\left\langle \frac{d_{vw} - c_{k,0}}{d - c_{k,0}}\right\rangle + \rho_{vfb} d_{vfb}\left\langle \frac{d_{vfb} - c_{k,0}}{d - c_{k,0}}\right\rangle\right)}{\lambda f_{cd}}$$

where $\langle x \rangle : (< 0 = 0; \ge 0 = x)$

and $\rho_{vij} = A_{vij} / \left(b_{sr0} d_{vij}\right)$ and $b_{sr0} = 1.0\,m$ is the unit moment strip.

Assuming that the shear key is positioned in the geometrical centre of the cross-section with

$$d_a = d_{aw} = h / 2 = 115\,mm$$

Considering the geometrical characteristics of the European standard HEB120 profile:

$$b_a = 120\,mm \quad h_a = 120\,mm \quad t_f = 11\,mm \quad t_w = 6.5\,mm \quad A_a = 3400\,mm^2$$

Check whether the shear key depth-to-effective bending depth is satisfied:

$$h_a / d = 120 / 198 = 0.60 \ge 0.43$$

It results in that the centroids of the shear key components are located at:

$$d_{aft} = d_a + h_a / 2 - t_f / 2 = 169.5\,mm$$

$$d_{afb} = d_a - h_a / 2 + t_f / 2 = 60.5\,mm$$

$$d_{aw} = d_a = 107.5\,mm$$

The corresponding reinforcing areas for the top flange, web and bottom flange are:

$$A_{aft} = A_{afb} = b_a \times t_f = 1320\,mm^2$$

$$A_{aw} = A_a - \left(A_{aft} + A_{afb}\right) = 1320\,mm^2$$

Following the iterative procedure, the depth of the compression zone in the composite region is obtained as:

$$c_k = 2.04 \times 33.8\,mm = 69.1\,mm \text{ with } \lambda_c = 2.04$$

Hence, the required section moment of inertia of the shear key is:

$$I_a \ge \frac{m_{Ed,max}\left(d - c_k\right)}{10 f_{ysd}} = \frac{78.2k\,Nm / m\left(198\,mm - 69.1\,mm\right)}{10 \times 435\,N / mm^2} = 231 \times 10^4\,mm^4$$

The section moment of inertial of the HEB120 profile is:

$$I_a = 864.4 \times 10^4 \, mm^4$$

Embedment length:

The embedment length l_v of one shear key may be calculated from the assumption that the critical section is situated at the composite-to-concrete interface. The required control perimeter $b_{0,req}$ for the required design shear force V_{Ed} is:

$$b_0 \geq \dfrac{V_{Ed}}{\dfrac{0.6}{\gamma_c} k_{pb.kc0} \left(100 \rho_l f_{ck} \dfrac{d_{dg}}{r_s}\right)^{1/3}} = \dfrac{476 \, kN}{\dfrac{0.6}{1.5} 4.0 \left(100 \times 0.0063 \times 30 \, MPa \times \dfrac{36 \, mm}{1167 \, mm}\right)^{1/3}}$$

$$= 3564 \, mm$$

in which

$$d_{dg} = 16 \, mm + d_g = 36 \, mm$$

$k_{pb,kc,0} = 4.0$ (value which can be considered at the pre-sizing

$$m_{Rd,k} = f_{ysd} \left\{ \rho_l d \left(d - \dfrac{c_k}{2} \right) + \Sigma \left[\rho_{vij} d_{vij} \left\langle (d_{vij} - c_k)(d_{vij} - c_k / 2) \right\rangle \right] / (d - c_k) \right\}$$

stage)
$$= 435 \left\{ 0.0063 \times 198 \left(198 - \dfrac{69.1}{2} \right) + \Sigma (139 + 22 + 0) \right\}$$

$$= 435 (203 + 139 + 22 + 0) = 158.2 \, kNm \, / \, m$$

And d_0 is the effective shear depth

$$d_0 = d - (d_a - b_a / 2 + t_f) = 189 - (115 - 120 / 2 + 8) = 126.5 \, mm$$

The critical length which determines the location of the critical section is:

$$l_0 \geq \max \begin{vmatrix} (b_0 - \pi d_0) / 8 = (3564 - 126\pi) / 8 = 396 \, mm \\ (b_0 - \pi d - 2(b_{col} - b_a)\sqrt{2}) / (4\sqrt{2}) \\ = (3564 - 126\pi - 2(260 - 120)\sqrt{2}) / (4\sqrt{2}) = \\ = 450 \, mm \end{vmatrix}$$

$$\cong 450 \, mm$$

which leads to the required embedded length of the shear key:

$$l_v = l_0 - d_0 / 2 = 450 - 126 / 2 = 387.5 \cong 390\,mm \geq 2h_a = 240\,mm \geq d / 2$$

$$= 99\,mm$$

Punching shear design

$$V_{Rd,c} = \frac{0.6}{\gamma_c} k_{pb,kc} \left(100 \rho_l f_{ck} \frac{d_{dg}}{r_s} \right)^{1/3} b_0 d_0$$

in which

$$k_{pb,kc} = \sqrt{35\mu_p \frac{d}{b_0} \left(1 + 4 \frac{b_a}{b_{col}} \frac{l_v}{r_s} \left(\frac{m_{Rd,k}}{m_{Rd,c}} - 1 \right) \right)} = 4.77$$

The effective critical length is:

$$l_0 = l_v + d_0 / 2 = 390 + 126.5 / 2 = 453\,mm$$

The control perimeter is:

$$b_0 = \min \begin{cases} \pi d_0 + 8 l_0 = \pi 126 + 8 \times 453 = 4023\,mm \\ \pi d_0 + 4 \left[l_0 + \left(b_{col} - b_a \right) / 2 \right] \sqrt{2} = \pi 126 + 4\sqrt{2} \left(453 + 70 \right) = 3357\,mm \end{cases} = 3357\,mm$$

The punching shear strength is:

$$V_{Rd,c} = \frac{0.6}{\gamma_c} k_{pb,kc} \left(100 \rho_l f_{ck} \frac{d_{dg}}{r_s} \right)^{1/3} b_0 d_0 = \frac{0.6}{1.5} 4.77 \left(100 \times 0.0063 \times 30 \frac{36}{1167} \right) 3357 \times 126.5$$

$$= 677\,kN$$

$$V_{Rd.kc} < \frac{0.6}{\gamma_c} \sqrt{f_{ck}} b_0 d_0 = 930\,kN$$

Flexural strength of the hybrid connection

$$V_{flex} = \mu_p m_{Rd,c} \left[1 + 4 \frac{b_a}{b_{col}} \frac{l_v}{r_s} \left(\frac{m_{Rd,k}}{m_{Rd,c}} - 1 \right) \right] = 8 \times 98 \left[1 + 4 \frac{120}{260} \frac{390}{1167} \left(\frac{158}{98} - 1 \right) \right] = 1081\,kN$$

$$V_{flex} = 1081\,kN \geq V_{Ed} = 476\,kN$$

Shear key verification

$$V_{pl,a,Rd} = A_{a,v} \frac{f_{yak}}{\sqrt{3}} = 224 \, kN >$$

$$\frac{V_{a,Ed}}{V_{pl,a,Rd}} = 0.53 > 0.5 \text{-> the yield strength for assessment of bending capac-}$$

ity should be reduced by *(1-ρ)*.

$$\rho = \left(\frac{2V_{a,Ed}}{V_{pl,a,Rd}} - 1 \right)^2 = 0.0035$$

The force distribution coefficient for shear keys in flat slabs is determined below.

$$\kappa = \frac{1}{3} \left(1 - \frac{l_v}{r_s} \right) = 0.22$$

Using the assumed force distribution, the bending moment acting on a shear key is given by:

$$M_{a,Ed} = V_{a,Ed} \left[\kappa \frac{b_a}{8} + (1 - \kappa) \left(\frac{l_v}{2} + \frac{b_a}{8} \right) \right] = 19.84 \, kNm$$

Assuming that in flat slabs, the full plastic capacity can be achieved, the plastic section modulus of the HEB120 cross-section is:

$$W_{a,pl} = 165.2 \times 10^3 \, mm^2$$

The stiffness coefficient of the shear key is given by:

$$k_m = \left(65 \rho_l \right)^{\frac{1}{3}} \frac{d_0}{b_a} \left(\frac{l_v}{r_s} \right)^{\frac{2}{3}} = 0.38$$

The moment capacity of the shear key is:

$$M_{a,Rd} = k_m W_{a,pl} (1 - \rho) f_{yad} = 22.01 \, kNm$$

Ultimately, the bottom flange of the shear key should be wide enough to support the force transfer:

$$b_{a,min} = V_{a,Ed} \frac{2}{v f_{cd} l_v} = 61.0 \, mm < b_a = 120 \, mm, \text{ where } v = 0.5$$

The above results in a cruciform shear head with HEB120 cross-section and l_v=390 *mm* embedment length, which are adequate to transfer

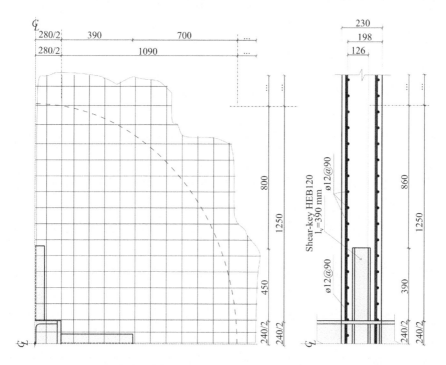

Figure 6.22 Designed hybrid connection to flat slabs without transverse reinforcement

the design forces. A detail of the designed hybrid connection is shown in Figure 6.22.

Transverse reinforcement design

This short example follows up the design of hybrid connections of steel columns to flat slabs by beams of cruciform shear-head systems. It is assumed that the design shear at the interior connection is V_{Ed}=1180 kN, whilst the bending and torsional moments are representative for the loading case giving the design shear force above. The purpose of this example is only to highlight the design procedures for hybrid connections provided with headed shear studs as transverse reinforcement whilst using the same flat slab details designed above. Note that the flexural reinforcement ratio in the example above may be too low to mobilize a punching shear failure and to activate the shear reinforcement. The flexural strength of the connection V_{flex}=1081 kN, assessed above, is below the design shear V_{Ed}=1180 kN. Hence, an increase in the amount of bending reinforcement would be required.

Assuming that the shear resistance of the hybrid connection without transverse reinforcement is taken from the previous example as $V_{Rd,c}$=677 kN

$$V_{Rd,c} = \frac{0.6}{\gamma_c} k_{pb,kc} \left(100\rho_l f_{ck} \frac{d_{dg}}{r_s} \right)^{1/3} b_0 d_0 = 677\,kN$$

The amount of transverse reinforcement required for shear assessments can be determined using the following relationship:

$$V_{Rd,s} = \frac{V_{Ed} - \eta_c V_{Rd,c}}{\eta_s} = \frac{1180 - 0.57 \times 677}{0.8} = 990 \, kN$$

Where:

$$\eta_c = V_{Rd,kc}/V_{Ed} = 0.57$$

$$\eta_s = \left(\frac{r_s}{d}\right)^{1/2} \left(\frac{0.8}{\eta_c \sqrt{8\mu_p d/b_0}}\right)^{3/2} = \left(\frac{1167}{198}\right)^{1/2} \left(\frac{0.8}{0.58\sqrt{8 \times 8 \times 198/3357}}\right)^{3/2} = 1.46 \leq 0.8$$

Considering that the headed shear studs are 10 mm in diameter and of B500S steel grade $f_{ywk} = 500 \, MPa \, \gamma_s = 1.15 f_{ywd} = 435 \, MPa$

$$A_{sw} = \frac{V_{Rd,s}}{f_{ywd}} = 2277 \, mm^2$$

It results in that the required amount of transverse bars n_{bw} located in the shear key region is:

$$n_{bw,req1} = \frac{A_{sw}}{\pi d_{bw}^2 / 4} = \frac{2277 \, mm^2}{78.5 \, mm^2} = 29.0 \cong 30 \, bars$$

The active transverse bars are those located in the vicinity of the shear key flanges. Hence, the required amount of bars has to be evenly distributed in the region of the flanges.

The maximum distance between the column face and the first headed shear stud is:

$$0.35 d_0 \leq s_0 \leq 0.75 d_0 \rightarrow 44 \, mm \leq s_0 \leq 95 \, mm \rightarrow s_0 = 70 \, mm$$

The distance between the edge of the flange and the row of studs is also limited to

$$0.3 d_0 \leq s_{0k} \leq 0.7 d_0 \rightarrow 37 \, mm \leq s_{0k} \leq 89 \rightarrow s_{0k} = 70 \, mm$$

The maximum distance between headed shear studs is the minimum between:

$$s_1 = \min(\leq 0.75 d_0, 300 \, mm) = \min(88, 300 \, mm) = 85 \, mm$$

It results in that the first headed shear stud is located at s_0=70 mm from the column face and the subsequent transverse bars at a spacing of s_1=85 mm following the direction of the flanges. They are parallel with the shear key flanges at s_{0k}=70 mm.

Based on the constructive shear reinforcement detailing and considering that the reinforcement has to be symmetrically arranged to the centre of the shear key on both sides, the amount of transverse bars required along on shear key edge is obtained as:

$$n_{bw,req2/side} = \frac{l_v - s_0}{s_1} = \frac{390 - 70}{100} = 3.2 \cong 4 \; bars$$

Whereas the total amount is:

$$n_{bw,req2} = 2 \times n_v \times n_{bw,req2/side} = 2 \times 4 \times 4 = 32 \; bars$$

The effective, and active, shear reinforcement is:

$$n_{bw} = 32 \; that \; \Sigma A_{sw} = 32 \times 78.5 \, mm^2 = 2513 \, mm^2 \geq A_{sw}$$

Supplementary "non-active" shear reinforcement with the same characteristics must be added at s_{0k}=80 mm from the tip of the shear key, but not further than s_1=100 mm from the most adjacent active shear reinforcement. Consequently, an amount of three shear-headed studs are added at the interface of each shear key. The total amount of non-active bars is 12.

To avoid punching outside of the shear-reinforced region, the slab has to be provided with reinforcement with the condition that the maximum in-plane distance between two reinforcement bars must be below $s_{out} \leq 3d_0 = 3 \times 126 = 380 \, mm$. To ensure that punching outside the shear-reinforced region is avoided, four reinforcement bars are added to each quarter of the slab following its diagonals.

Based on the previous treatment, the contribution of the shear reinforcement to punching shear capacity is:

$$V_{Rd,s} = A_{sw} f_{ywd} = 2513 \times 435 = 1093 \, kN$$

whereas the punching shear capacity of the flat slab is:

$$V_{Rd.cs} = \eta_c V_{Rd,c} + \eta_s V_{Rd,s} = 393 + 874 = 1262 \, kN > V_{Ed} = 1180 \, kN$$

Maximum punching shear capacity

$$V_{Rd,max} = \eta_{sys} V_{Rd,kc} = 1.8 \times 677 \, kN = 1218 \, kN$$

in which $\eta_{sys} = 1.8$ for double-headed shear studs

As observed, crushing of concrete does not govern the design ($V_{Rd,max}$=1226 kN > V_{Ed}=1180 kN), hence the amount of transverse reinforcement, shear key length and concrete strengths are adequate.

A detail of the transversely reinforced hybrid connection is shown in Figure 6.23, whilst a summary of the design results is listed in Table 6.2.

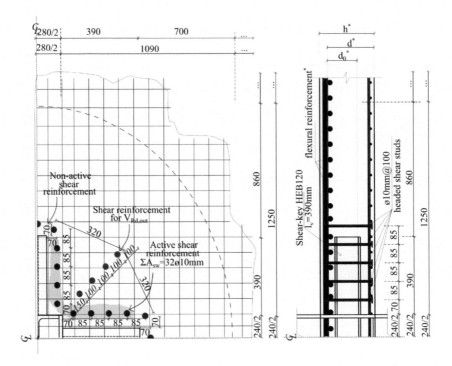

Figure 6.23 Designed hybrid connection to flat slabs with transverse reinforcement

Table 6.2 Results of the design example for the hybrid connection to flat slabs

Verification	Region	Cross-section	Capacity ($_{Rd,i}$)		Demand ($_{Ed,i}$)	($_{Rd,i}/_{Ed,i}$)
Bending	Composite	Composite	158.2	>	78.2	2.02
Flexural capacity	Composite	Composite	1081	>	476	2.27
Bending	Composite	Shear key	22.0	>	19.9	1.11
Shear	Composite	Shear key	224	>	119	1.88
Bending	Transition	Reinforced concrete	98.0	>	78.2	1.25
Punching shear	Transition	Reinforced concrete	677	>	476	1.42
Punching shear limit	Transition	Reinforced concrete	930	>	476	1.95
Punching shear with headed studs	Transition	Reinforced concrete	1262	>	1180	1.07
Maximum punching shear capacity	Transition	Reinforced concrete	1218	>	1180	1.03

LIST OF NOTATIONS

Lowercase Latin letters

a_v:	shear span-to-depth ratios around
b_0:	critical perimeter
b_a:	shear key width
b_c:	width of the beam component
$b_{c,init}$:	initial beam width
b_{c1}, b_{c2}:	column sizes
b_{col}:	steel columns size
b_s:	active strip width
b_{sr0}:	the unit moment strip
c_c:	depth of the compression zone in a reinforced cross-section
c_k:	depth of the compression zone in a composite cross-section
$c_{k,pl}$:	compression zone depth assessed using a plastic distribution in the cross-section
$c_{min,dur}$:	minimum concrete cover due to durability requirements
$c_{min,sl}$:	minimum cover required for longitudinal bars
$c_{min,sw}$:	minimum cover required for transverse bars
c_{nom}:	nominal concrete cover
c_z:	concrete cover
d:	effective bending depth to the centroid of longitudinal reinforcement
d_0:	effective shear depth
d_a:	effective bending depth of the shear key assessed from the outermost compression fibre and centroid of the steel insert
d_{afb}:	effective bending depth to the centroid of the bottom flange
d_{aft}:	effective bending depth to the centroid of the top flange
d_{aij}:	effective bending depth of shear key component (flange or web)
d_{aw}:	effective bending depth to the centroid of the web
d_{ba}:	diameter of the welded shear stud
d_{bl}:	diameter of the longitudinal reinforcement bars
$d_{bl,max}$:	maximum diameter of the longitudinal reinforcement bars
d_{bw}:	diameter of the transverse reinforcement bars
d_{dg}:	reference value of the roughness of the critical shear crack
d_g:	maximum aggregate size
d_{g0}:	the reference aggregate size
d_{init}:	initial effective bending strength
f_c:	test cylinder concrete compressive strength
f_{cd}:	design concrete compressive strength
f_{ck}:	characteristic concrete compressive strength
f_u:	ultimate tensile strength of the steel in welded shear studs
f_{yad}:	design yielding strength of the shear key
f_{yak}:	characteristic yield strength of the steel in the shear key

f_{ysd}:　design yield strength of the longitudinal reinforcement
f_{ysd}:　design yield strength of the longitudinal reinforcement
f_{ysk}:　characteristic yield strength of the longitudinal reinforcement
f_{ywd}:　design yield strength of the transverse reinforcement
$f_{ywd,}$:　yield strength of the shear reinforcement
f_{ywk}:　characteristic yield strength of the transverse reinforcement
h_a:　shear key depth
h_{init}:　initial beam depth
$h_{s,init}$:　initial flat slab depth
k_m:　reduction factor of the plastic moment capacity of the shear key for hybrid flat slab connections
$k_{pb,kc}$:　parameter for assessment of the punching shear strength
l_0:　critical length
l_{cr}:　radial crack length
l_v:　shear key embedment length
$l_{v,i}$:　shear key embedment length
$l_{v,init}$:　initial shear key embedment length
m_{Ed}:　design bending moment per unit length
$m_{Rd,c}$:　plastic moment resistance per unit width of a reinforced concrete cross-section
$m_{Rd,k}$:　plastic moment resistance per unit width of a composite cross-section
n_{bl}:　number of longitudinal reinforcement bars
n_{bw}:　number of transverse bars in a section
n_v:　number of shear keys in a shear head
p_0:　average allowed drilling distance in the column flanges
r:　fillet radius
r_s:　the distance between the column centre and zero bending
$s_{l,min}$:　minimum spacing between longitudinal bars
$s_{t,max}$:　maximum distance between transverse reinforcement bars
$s_{t,nc}$:　transverse reinforcement spacing in the non-composite region
t_f:　shear key flange thickness
t_w:　shear key web thickness
z_c:　bending lever arm to the longitudinal reinforcement in the non-composite region
z_k:　bending lever arm to the longitudinal reinforcement in the composite region
z_{kc}:　bending lever arm to the longitudinal reinforcement in the transition region

Uppercase Latin letters

A_a:　shear key total cross-sectional area
$A_{a,v}$:　shear area of the shear key

A_{afb}:	cross-sectional area of the bottom flange
A_{aft}:	cross-sectional area of the top flange
A_{aij}:	area of a shear key component (flange or web)
A_{aw}:	cross-sectional area of the web
A_{dbw}:	area of a transverse reinforcement bar
A_{sl}:	longitudinal reinforcement area
A_{sw}:	area of transverse reinforcement
A_{sw}:	transverse reinforcement area
A_{vL}:	the contact area of the steel profile
E_a:	shear key steel elastic modulus
E_c:	concrete elastic modulus
I_a:	second moment of inertia of the shear key cross-section
I_c:	second moment of inertia of the concrete cross-section
L, L_s, L_y, L_x:	span
$M_{a,Ed}$:	design moment acting on the shear key
$M_{a,el,Rd}$:	elastic bending moment resistance
$M_{a,pl,Rd}$:	plastic bending moment resistance
M_{Ed}:	design bending moment
$M_{Ed,c}$:	design bending moment in a reinforced concrete cross-section in the non-composite region
$M_{Ed,k}$:	design bending moment for a composite cross-section in the composite region
$M_{Ed,kc}$:	design bending moment for a composite cross-section in the transition region
$M_{Ed,k}$:	design bending moment in the composite region
M_{Rc}:	design bending moment resistance of the steel column
M_{Rb}:	design bending moment resistance of the beam at the reinforced concrete cross-section
$M_{Rd,c}$:	design bending moment resistance of a reinforced concrete cross-section
$M_{Rd,k}$:	design bending moment resistance of a composite cross-section
P_{flex}:	component bending strength
P_{Rd}:	resistance of a welded shear stud
P_{test}:	component test strength
$V_{a,Ed}$:	design shear force acting on the shear key
V_{calc}:	assessed strength
V_{Ed}:	design shear or punching shear force
$V_{Ed,c}$:	design shear force in the non-composite region
$V_{Ed,k}$:	design shear force in the composite region
$V_{Ed,kc}$:	design shear force in the transition region
V_{flex}:	flexural strength
$V_{pl,a,Rd}$:	plastic shear capacity of the shear key
V_{Rd}:	design shear or punching shear resistance

$V_{Rd,c}$:   concrete shear capacity or contribution of concrete to the shear or punching shear capacity

$V_{Rd,c}$:   contribution of the concrete to the punching shear capacity

$V_{Rd,c,kc}$:   shear resistance of components not requiring shear reinforcement in the transition region

$V_{Rd,cs,in}$:   punching shear capacity assuming failure within the shear-reinforced region

$V_{Rd,cs,out}$:   punching shear capacity assuming failure outside the shear-reinforced region

$V_{Rd,k}$:   shear resistance in the composite region

$V_{Rd,L,k}$:   longitudinal shear strength

$V_{Rd,max}$:   maximum punching shear capacity of shear-reinforced connections

$V_{Rd,s}$:   contribution of shear reinforcement to the punching shear capacity

$V_{Rd,s,k}$:   contribution of shear reinforcement to the shear resistance in the composite region

V_{test}:   test shear capacity

$V_{test/num}$:   test or numerical shear strength

$W_{el,a}$:   shear key elastic section modulus

$W_{pl,a}$:   shear key plastic section modulus

Greek letters

β_c: concrete cover adjustment factor

γ_a: partial safety factor for welded shear studs

γ_c: concrete safety factor

γ_i: material safety factors

γ_{M0}:   shear key steel partial safety factor

γ_{Rd}:   overstrength factor

γ_s: reinforcement steel partial safety factor

η_c: factor that captures the concrete contribution at critical crack development

η_s: factor that accounts for the level of activation of the transverse bars

θ: critical shear crack inclination angle

λ: stress distribution in the compression zone

λ_c: proportionality factor

λ_β: shear resistance enhancement factor

μ_p: factor for the type of column to flat slab connection

ν: compressive strength reduction

$\nu_{Ed,avg}$:   average reaction force

ν_{Ed1}, ν_{Ed2}:   reaction forces

$\nu_{perp,max}$:   maximum perpendicular shear force per unit width

ρ: yield strength reduction factor for bending resistance

ρ_a: shear key reinforcement ratio

ρ_{afb}:   flexural reinforcement ratio of the bottom shear key flange
ρ_{aft}:   flexural reinforcement ratio of the top shear key flange
ρ_{aw}:   flexural reinforcement ratio of the shear key web
ρ_{l}:   flexural reinforcement ratio of a concrete section
ρ_{tot}:   combined reinforcement ratio
σ_{sl}:   stress in the longitudinal reinforcement
τ_{Rd}:   design shear strength for longitudinal shear
ψ: slab rotation
κ: force distribution factor along the shear key

REFERENCES

Bompa, D. V., and Elghazouli, A. Y. (2015) Ultimate shear behaviour of hybrid reinforced concrete beam-to-steel column assemblages. *Engineering Structures*, 101, 318–336. https://doi.org/10.1016/j.engstruct.2015.07.033

Bompa, D. V., and Elghazouli, A. Y. (2016) Structural performance of RC flat slabs connected to steel columns with shear heads. *Engineering Structures*, 117, 161–183. https://doi.org/10.1016/j.engstruct.2016.03.022

Bompa, D. V., and Elghazouli, A. Y. (2017) Numerical modelling and parametric assessment of hybrid flat slabs with steel shear heads. *Engineering Structures*, 142, 67–83. https://doi.org/10.1016/j.engstruct.2017.03.070

Bompa, D. V., and Elghazouli, A. Y. (2020) Nonlinear numerical simulation of punching shear behavior of reinforced concrete flat slabs with shear-heads. *Frontiers of Structural and Civil Engineering*, 14, 3310356. https://doi.org/10.1007/s11709-019-0596-5

CEN (European Committee for Standardization). (2004a) Eurocode 4: Design of composite steel and concrete structures – Part 1–1: General rules for buildings. Brussels.

CEN (European Committee for Standardization). (2004b) Eurocode 2: Design of concrete structures – Part 1–1: General rules and rules for buildings. Brussels.

CEN (European Committee for Standardization). (2004c) Eurocode 8: Design of structures for earthquake resistance – Part 1: General rules, seismic actions and rules for buildings. Brussels.

CEN (European Committee for Standardization). (2005) Eurocode 3: Design of steel structures – Part 1–1: General rules and rules for buildings. Brussels.

CEN (European Committee for Standardization). (2018) Eurocode 2: Design of concrete structures—General rules and rules for buildings, final version of PT1-draft prEN 1992-1-1:2018. Brussels.

Eder, M. A., Vollum, R. L., and Elghazouli, A. Y. (2011) Inelastic behaviour of tubular column-to-flat slab connections. *Journal of Constructional Steel Research*, 67(7), 1164–1173. https://doi.org/10.1016/j.jcsr.2011.02.009

Eder, M. A., Vollum, R. L., and Elghazouli, A. Y. (2012). Performance of ductile RC flat slab to steel column connections under cyclic loading. *Engineering Structures*, 36, 239–257. https://doi.org/10.1016/j.engstruct.2011.12.002

fib (Fédération Internationale du Béton). (2012) Model code 2010 - final draft, vol. 1, and vol. 2, fib Bulletins 65, and 66, 2012 Lausanne, Switzerland.

Guandalini, S., Burdet, O., and Muttoni, A. (2009) Punching tests of slabs with low reinforcement ratios. *ACI Structural Journal*, 106(1), 87–95.

Kim, J. W., Lee, C. H., and Kang, T. H. K. (2014) Shearhead reinforcement for concrete slab to concrete-filled tube column connections. *ACI Structural Journal*, 111(3), 629–638.

Lee, C. H., Kang, T. H. K., Kim, J. W., Song, J. K., and Kim, S. (2019). Seismic performance of concrete-filled tube column-reinforced concrete slab connections with shearhead keys. *ACI Structural Journal*, 116(2), 233–244.

Lee, C. H., Kim, J. W., and Song, J. G. (2008) Punching shear strength and post-punching behavior of CFT column to RC flat plate connections. *Journal of Constructional Steel Research*, 64(4), 418–428. https://doi.org/10.1016/j.jcsr.2007.08.003

Moharram, M. I., Bompa, D. V., and Elghazouli, A. Y. (2017a) Experimental and numerical assessment of mixed RC beam and steel column systems. *Journal of Constructional Steel Research*, 131, 51–67. https://doi.org/10.1016/j.jcsr.2016.12.019

Moharram, M. I., Bompa, D. V., and Elghazouli, A. Y. (2017b) Performance and design of shear-keys in hybrid RC beam and steel column systems. *ce/Papers*, 1(2–3), 2031–2040. https://doi.org/10.1002/cepa.248

Moharram, M. I., Bompa, D. V., Xu, B., and Elghazouli, A. Y. (2022) Behaviour and design of hybrid RC beam-to-steel column connections. *Engineering Structures*, 250, 113502. https://doi.org/10.1016/j.engstruct.2021.113502

Moharram, M. I. S. (2018) *Inelastic behaviour of hybrid reinforced concrete beam and steel column systems*. PhD Dissertation. Imperial College London, Department of Civil and Environmental Engineering. https://doi.org/10.25560/72837

Muttoni, A. (2008) Punching shear strength of reinforced concrete slabs without transverse reinforcement. *ACI Structural Journal*, 105(4), 440–450.

Muttoni, A., Fernández Ruiz, M., and Simões, J. T. (2018) The theoretical principles of the critical shear crack theory for punching shear failures and derivation of consistent closed-form design expressions. *Structural Concrete*, 19(1), 174–190. https://doi.org/10.1002/suco.201700088

Vecchio, F. J., and Shim W. (2004) Experimental and analytical reexamination of classic concrete beam tests. *Journal of Structural Engineering*, 130(9), 460–469. https://doi.org/10.1061/(ASCE)0733-9445(2004)130:3(460)

Index

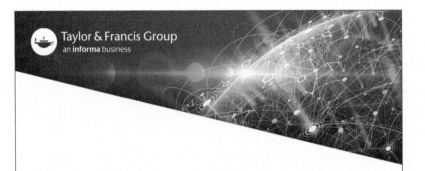